WHAT PEOPLE
ABOUT TH

"I'm only a few chapters in but am already finding myself willing you on! The book is beautifully written, with a real sense of movement as you naturally flow from one part of the story to the next. You write thoughtfully and with such humility. I am already reflecting on aspects of my own life as a result, and thinking more about how I might find my own purpose and meaning." - **Lou**

"What a great read. This book is engaging, interesting, fascinating, inspiring, honest and emotional. Part memoir, part travelogue and great storytelling. I cried as you described your childhood. Many of your struggles resonated with my own experiences. Thank you for sharing your story so openly with us." - **Emma**

"Ken is a born storyteller. In his book he comes across as sincere, honest, open and human. His is a story told without self-pity and without arrogance, and boy does he have a story to tell. Ken's writing is incredibly engaging and I found myself unable to stop. I also love the way the book seamlessly incorporates little bits of history as it takes you around the world. A real joy to read." - **Alison**

"A natural storyteller, Ken weaves his way seamlessly through his many life experiences. *The Pursuit of Purpose* is jam-packed with refreshing honesty, real sadness and more than a little joyful optimism, all sprinkled with many fascinating insights. As a writer, he shows a huge dollop of humanity, with a clear love of people and how we all relate to our precarious world." - **Richard**

"This book is written in such a vivid way. I felt like I was sitting on Ken's shoulder, travelling with him as he seeks out his purpose, seeing and feeling and experiencing every stage of his journey. The biggest takeaway for me? Go through life with open eyes, take chances, trust people, and stay curious." - **Sebastian**

WHAT PEOPLE ARE SAYING
ABOUT THIS BOOK

"Ken has led a remarkable life. He's brushed shoulders with some of the greatest figures of our generation, and started a movement that would revolutionize the way that millions of people communicate with each other. His book is a fascinating journey through the many lives he's led, the people that have shaped him along the way, and the unexpected twists and turns of life that nobody can anticipate. This is the most personal work that Ken has ever published. In the intimate details of this testimony, I've been inspired to learn this truth: that ordinary people really can lead extraordinary lives." - **Rowena**

"This book comes across with humility, and in an engaging, relatable voice. Just part of the way through I felt inspired to pursue my own path of purpose. Anyone who picks up Ken's book will find it easy to resonate with the fluctuating emotions of life. It's a great example of sharing experiences to help others along the way. A brilliant read." - **Samantha**

"Ken offers a front row seat to a fascinating journey of self-discovery written with integrity, passion, self-effacing humour and humility. Without a clear roadmap he has pursued his passions with remarkable commitment and a generous spirit. His reward has been extraordinary experiences and a resolution to his own question of purpose. Anyone who has not yet found an answer will surely double-down on the effort after reading this book." - **Roger**

"*The Pursuit of Purpose* is a collection of stories about adventure, strength, resilience and multiple new beginnings, but also about self-doubt, insecurity and uncertainty. If you believe you are made for more you will feel connected with Ken's story, and encouraged and validated in your own search for purpose. This book, written so beautifully, will capture your attention and inspire you to embrace your journey with hope. A wonderful, enlightening read." - **Lulu**

WHAT PEOPLE ARE SAYING
ABOUT THIS BOOK

"From developing a messaging platform that helped monitor elections in Nigeria, to running primate sanctuaries, to working with the likes of Archbishop Desmond Tutu, Ken has spent his life enriching the world instead of himself. In this honest and revealing book, he shares his journey and most intimate thoughts in the hope that it nudges us all to seize every opportunity to make a positive difference to the lives of others." - **Cathy**

"Ken tells his story with such modesty, from his childhood under relatively difficult circumstances on a housing estate in Jersey all the way to his adventures across Africa. There he provides a means for people to communicate easily with each other, bringing him into contact with a number of well-known figures. A fascinating story leading to fulfilment told in a very engaging and readable way." - **Roger**

"I was thrilled to be offered the chance to read an early copy of *The Pursuit of Purpose*, but at the same time was unusually nervous. I've long been an admirer of Ken and his work, but still felt I only knew half the story. As it turns out, I knew even less than that. His memoir is written beautifully, with real humility, and my admiration for his achievements from the humblest of beginnings has only grown. Hats off to you, Ken. I hope your book inspires others to pursue a life of meaning – and purpose." - **Carl**

THE PURSUIT OF PURPOSE

Written by

Ken Banks

Published by kiwanja.net

Published by kiwanja.net
www.kiwanja.net

ISBN: 978-1-7397717-0-6 (paperback)

First imprint 2022

A catalogue record for this book is
available from the British Library

This book has been composed in Playfair Display
by Luke Richards Design

Cover image artwork by Monon Mahfuz Saad

Printed in the United Kingdom by
Book Printing UK, Peterborough

Also by Ken Banks

The Rise of the Reluctant Innovator

Social Entrepreneurship and Innovation:
International Case Studies and Practice

Musings of a Mobile Anthropologist

*In memory of a wonderful
and supportive mother, and for
Henry, Maddie and Ollie, whom she
sadly never got to meet.*

*To my wife, Elina, for sticking
with me through thick and thin.*

*And to the late Archbishop Desmond
Tutu, for the most humbling – and
unlikeliest – of friendships.*

Contents

Part III: Purpose Explored

THE PURSUIT OF PURPOSE

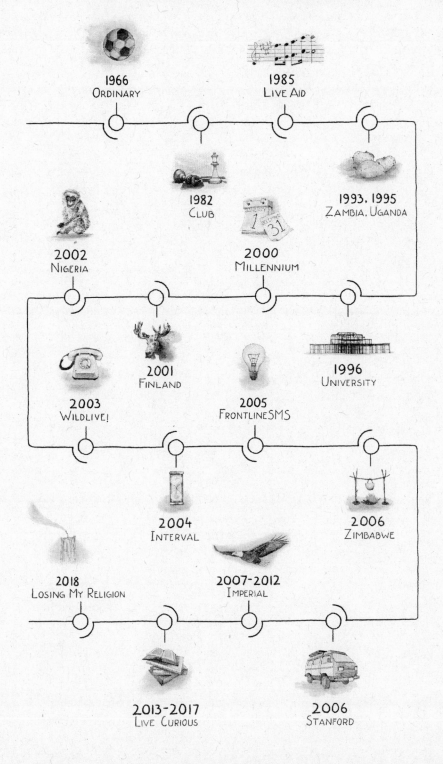

The Pursuit of Purpose

Part Memoir, Part Study – A Book About
Finding Your Way in the World

Acknowledgements

My decision to self-publish this book was an easy one, even if it did mean all of the hard work would fall on my own shoulders. What I'd lack in professional marketing and publishing expertise I'd gain in creative freedom, and I felt creative freedom with a book which has become incredibly personal to me was the right way – probably the only way – to go.

I could easily have filled a few pages here thanking everyone I met and worked with over the years, but nobody would read it all and most of you know who you are anyway. Instead, I'd like to thank those who offered up their skills and expertise as I grappled with the immense personal challenge of writing and publishing this book. A huge thank you, then, to all the artists, voiceovers, editors, typesetters, reviewers, web developers and financial backers, without whom the book would have fallen well short. Memoirs are not for the faint-hearted, and they're the sort of book you should only attempt once.

Among those I've drawn on to bring this book to life are Roger Fenner (who helped unearth the richest of family histories), Greg Watts (creative writing advice), Monon Mahfuz Saad (for all the beautiful hand-drawn pictures), Alison Bosman (copy editing and no-nonsense advice), Luke Richards (typesetting), Roger and Debbie Klene (early financial support and unwavering belief), Shane Casey and Omar Khalil (audiobook samples), Mino Stojanoski (promotional video producer), Bohdan Berezenko (ancestry research), Samantha Long and Deepak Goyal (website design and creation), Biljana K Tasetovikj (academic research), all 155 backers on Kickstarter (funding), and everyone who reviewed early drafts and gave valuable feedback, some of which you'll find on the 'What People Are Saying About This Book' pages.

To all of you, thank you from the bottom of my heart. Give yourselves a hug.

About the Author

With a career in global conservation and development spanning over two decades, Ken Banks is an award-winning founder, technologist, anthropologist and author committed to supporting positive social and environmental change around the world. His varied career has been shaped largely by contributions to the worlds of social entrepreneurship and social innovation, and the support of conservation and development projects around the world, with a particular focus on mobile technology in Africa.

Following an early career developing financial and accounting systems from his home in Jersey, Ken has since received global recognition in his field thanks to his pioneering work developing a text messaging platform called FrontlineSMS, which scaled to over 190 countries and benefited tens of millions of people. He is also well known for his wider speaking, writing and mentoring efforts.

He is a PopTech Fellow, a Tech Awards Laureate, an Ashoka Fellow and a National Geographic Explorer. In 2013 he was nominated for the TED Prize. In recognition of his pioneering work, he was presented with the 2017 Eugene L Lawler Award for Humanitarian Contributions within Computer Science by the Association for Computing Machinery in the USA, and in 2018 was appointed a Visiting Fellow at the prestigious Judge Business School in Cambridge.

He has written widely on technology and social innovation, and his work has been published by the BBC, CNN, WIRED and *The Guardian*, among many others. Ken is also the editor of two books on social innovation and entrepreneurship, the first of which has a 5-star rating and became an Amazon 'Development Studies' bestseller. *The Pursuit of Purpose* is his fourth, and likely last, book.

A Book is Born

I must be honest, this is a book I never wanted to write. Frustration, self-doubt, unhealthy doses of imposter syndrome and frequent bouts of inferiority complex all combined to keep pen away from paper for years. When I was finally convinced that I should write it, I had no idea where, or how, to begin.

But then came the pandemic, multiple lockdowns, and more than enough time to pause and reflect. Increasingly curious children, a chorus of people telling me I needed to share my story, the unearthing of a rich family history and the personal need to reconcile myself with my own work and journey all, eventually, conspired against me. A story that I struggled to believe was mine to tell slowly began taking shape exactly two years ago amid a raging pandemic, and today *The Pursuit of Purpose* is the result.

This is a book about purpose and meaning, things so many of us seek at some point in our lives. It's about putting yourself out there whatever the personal or financial cost, about relentlessly grabbing every opportunity that comes your way, about never giving up and, perhaps most importantly, about never losing hope. It's also about serendipity and chance, and random acts such as travelling across Africa with the British Prime Minister, or convincing the likes of Archbishop Desmond Tutu, or musician Peter Gabriel, to write forewords for my first two books. It's about being in the right place at the right time – more than once – and those incredibly rare moments of insight that can lead to the development of simple solutions to big problems, solutions that help tens of millions of people around the world, and the awards I could never have dreamed of winning for being fortunate enough to do so.

It's a story of living my purpose not once, but twice as I found meaning in different settings. And it's a story of many others I met along the way – the living, and the dead.

In short, it's a book about starting with nothing in search of everything, and all the little surprises that life throws at you along the way. It's about playing the long game in the pursuit of purpose, and the incredible things that can happen when you're fortunate enough to find it.

This book is a story in three parts. *Purpose Pursued* begins on the beautiful island of Jersey, where a difficult early childhood shaped and influenced me. We learn how a global music event led to an unexpected awakening, and the lengths I went to try and find some sort of purpose and meaning in life. From there we travel to England, and to Zambia, Uganda, Nigeria, Cameroon and Finland where, time after time, I went out in search of answers that rarely came. Later, in *Purpose Lived*, we'll find out what happened when I eventually found it, and lived it, and the wider impact that finding purpose has had not just on me, but on many others I met and worked with along the way. Then, to wrap things up, *Purpose Explored* looks at the meaning of purpose, how and why it matters in different cultures, the role our ancestors might play in our own drive for purpose, and any lessons we might learn from my own particular journey. Depending on where you are in your life, or what interests you the most, it should be possible to read these parts in any order.

Putting these words to paper has been a big deal for me. Fortunately, the book took on a life of its own. Wherever the words ended up coming from, thank you. Without you I may never have finished.

But finish I did. My hope now is that this book somehow helps you, the reader, understand the ins and outs of your own journey and your own search for purpose, and that it gives you the courage to continue on, regardless of anything your own life may throw at you.

With peace, love – and purpose.

St Ives, Cambridgeshire
January 2022

Imposter

They say a picture paints a thousand words, but mine only tell me one thing.

'Ken, you really shouldn't be here.'

There was no way I could have known, but what happened that night on a busy highway in southern Nigeria was to change everything. Things were still a blur by the time I came to and found myself lying on my back in the middle of the road, the warm tarmac the only thing giving me comfort on an otherwise cool night. It wasn't until I tried to stand to get out of the oncoming traffic that I realised the true extent of my injuries, my foot turning sideways as I placed it awkwardly on the ground. I could see Chris, who had also been on the bike, miraculously running uninjured towards me through crowds of people and the flickering glare of lights, and could hear Jerry behind, screaming for help. Despite the pain, chaos and trauma, what happened to me in that split second turned out to be the defining moment of my life.

And now I sit, almost 20 years later, in a comfortable home I never expected to own, with a wonderful, healthy family I never expected to have, surrounded by dozens of reminders and mementos of a story I never thought would be mine to tell.

On one wall hangs a photo of me and Bill Clinton, smiling and shaking hands at a corporate event in Hong Kong. There's also one of me and Archbishop Desmond Tutu, sitting on a cruise ship in front of an auditorium packed with eager young students, sharing stories about how to change the world. Beside that is a framed copy of a page from the June 2010 edition of *National Geographic* magazine announcing my Explorer Award. And then – my children's favourite – a photo of Sir David Attenborough and me at an event in Bristol, where we both spoke about our passion for conservation. There's also a selection of crystal, glass, wooden and silicon trophies that were presented to me by my peers during a frantic five-year spell when my work was everywhere and it seemed I could do no wrong. And, as if that weren't enough, under my desk sits a large wooden box packed with magazines and newspapers that have featured my work over the years – *The Guardian, Economist, BBC Focus Magazine,* the *New York Times, Harvard Business Review* and the *Financial Times* among them. If I wasn't winning something, or being given money, I was flying to places I had never imagined flying, shaking hands with people I had never dreamed of meeting, or standing on a podium sharing the most unbelievable and unlikeliest of stories.

This book has been over 50 years in the making. It's about finding meaning, passion and purpose despite everything that life throws at you. It's about living a life you may not feel you deserve, and the emotions that can sometimes go with exceeding all of your own, and everyone else's, expectations. It's about insecurity and going nowhere for what feels like forever, but then ending up in places you'd only previously seen on the covers of travel magazines. It's about the people I met in these places, befriended and worked with, including activists, conservationists and humanitarians, all fighting for justice and a better world for everyone – a fight I felt compelled to join. But first and foremost it's about making the world a better place, a search for purpose that almost cost me my life, and a journey with the unlikeliest of beginnings on a run-down housing estate in Jersey.

It's also a story about family, and delving into the past in the hope of finding answers to the present. It's about discovering ancestors who found their own passion and purpose, driving them to travel widely and become mayors, successful authors, magistrates, journalists, inventors, composers, railway pioneers, saddlers to the Royal Family, and even codebreakers at Bletchley Park. It's about living up to the expectations of those who came before me, ancestors whose own stories help me understand what makes me who I am today.

Because the truth is that, after years of struggle, graft, frustration and self-doubt, I somehow believed myself unworthy of any of the success that did, eventually, come my way. And as the years have passed, coming to terms with any of it hasn't got any easier. Looking back on those successes I feel less of a sense of pride and more a feeling of 'How on earth did I get here?' and 'Why me?' It's strange, disturbing and surreal all at the same time, like some sort of out-of-body or out-of-mind experience that never really goes away.

Feelings like these are surprisingly common in anyone who finds themselves unable to accept their success is deserved, or that it's a direct result of their own skills or efforts. The creative industries, for example, are full of people who share their fame with a large serving of self-doubt. It's an industry where imposter syndrome (as it's better known) is almost considered a rite of passage. Even David Bowie struggled to reconcile himself with his music. 'I had enormous self-image problems and low self-esteem, which I hid behind obsessive writing and performing. I was driven to get through life very quickly. I really felt so utterly inadequate. I thought the work was the only thing of value.' For me, there's a certain authenticity and purity that comes with this sort of approach. The world seems to be full of reality TV stars, the sort of people who are somehow 'famous for being famous'. It's a strange world where egos dominate. Removing the person and the ego from the work gives it an opportunity to stand up in its own right, to shine and be scrutinised for what it is, not for who created it. There's a lot I like about that.

Perhaps, unconsciously, that kind of approach has been a draw for me. In the end, trying to become as invisible as possible became a large part of my imposter syndrome defence mechanism. For it to feel pure and genuine, it had to be all about the work and not about 'me'. My kiwanja.net website, for example, is largely written in the third person. The word 'I' is missing throughout, with the exception of the odd page here and there. The work is front and centre, and the person behind it is largely hidden away, playing little more than a supporting role. I am oddly absent from the vast majority of my life's work. Despite a 25-year career in the humanitarian sector there is, to my knowledge, only one photograph of me 'working in the field' taken a few years ago in Malawi during a CARE International trip. It's that photo, snapped in an empty, run-down classroom, with a faded blackboard and paint peeling off the walls, that reminds me of where I was happiest. A copy of that picture also hangs proudly on my wall.

Despite the runaway success of FrontlineSMS, a text messaging platform I developed for non-profit organisations in the developing world, there are no photographs of me with any of the people in the 190 or so countries where it ended up being used. I loved the fact that I was able to build something with a global reach, yet could remain largely out of sight – prize-winning or conference speeches aside, of course. In 2010 when I received my award from the National Geographic Society, their marketing team asked for photos of me with users, perhaps training them or doing a demonstration of the software. They couldn't figure out why I didn't have any. My answer – why would I? For me, the whole point was to develop something people could just take and use by themselves, on their own terms, and that's precisely what they managed to do. The reason it worked so well was because I *didn't* have to jump on a plane and help them. I've long believed that giving people this kind of responsibility is the purest form of empowerment. And I remain immensely proud, to this day, that many of those users achieved great things without my needing to guide, advise or help them in any way. All I did was create the platform, make it available to them, trust them, step back and let them get on with it. Oh, and share their stories if they wanted me to. How inspiring it was to share their stories.

If anything, trying to remain so completely invisible has probably contributed more to my feeling of self-doubt than I realise. But the alternative felt worse. The idea of sticking my name or my photograph everywhere, taking all the credit, making it all about 'me', giving off signals of self-importance, bragging, talking myself up and seeking worship all smelt of ego, and I've never been one for that. I'd choose humility every time. The obvious tension between remaining hidden and sharing my journey within the covers of this book is touched on a little later.

In 2008 I came across a book, by chance, that outlined an approach which strongly resonated with the one I had instinctively taken in my own work. *The Starfish and the Spider*, by Ori Brafman and Rod Beckstrom, explores the concept of leaderless organisations – organised communities, you might call them – where no single person is in charge, and there is no legal entity that holds everything together. I loved the idea of applying this approach to FrontlineSMS, to help build the most open of organisations, owned and run by the users, with no management structure, little to no overheads and no one really in charge. Given I was always more interested in scaling my impact than my organisation, it was a timely and inspiring read. The idea that I might be able to create an organisation without a CEO, at a time when I found myself meeting so many people who were desperate to become one, brought out the rebel in me. (For the record, I've always favoured the word 'Founder' which sounded, and felt, far more grounded, friendly and approachable. To the kinds of people I was beginning to work with, CEO just came across as a little too intimidating, a little too corporate, and a little too cold.)

Ambitions to tread this alternative path soon disappeared when I received a $1 million investment from Omidyar, the philanthropic firm created by the founder of eBay, Pierre Omidyar. The funds were given to help develop and grow my work. I remember those early conversations well, with the first taking place over burger and chips in The Baron of Beef, a traditional English pub close to the river in Cambridge town centre. Taking the money put an end to any ambitions I had of trying the braver, alternative 'starfish and spider' approach. Although we still ended up doing great things, I wonder, to this day, what might have happened if I'd stuck to my instincts. I guess I wasn't brave enough to turn down a million dollars. It was a lot of money, and a lot of faith, being put in someone who, not that long ago, had started with nothing. Other, 'smarter' people felt my work was worth investing in and that organisational scaling was the way to go. Who was I to argue?

Indeed, all of the CEOs, the PhDs, the experts and the global leaders I met felt just like that to me – smarter. Looking back, it's clear there was a touch of inferiority complex at play here. I rarely felt I deserved to be in the same room as many of them. This is a particularly unhelpful mindset if you've ever been hired as a consultant, as I have, given your primary role is often to be the most knowledgeable person in that room. On many occasions, being asked to deliver advice or ideas and answers on demand was very stressful for me, yet I somehow managed to do it well enough to be paid, thanked and asked back to do it again. I still feel as though I got away with a lot, despite the value that people claim my contributions made to their work.

I also frequently felt that I wasn't 'made of the right stuff' and that I wasn't of the right stock, particularly during my time in the USA where most people I met seemed to come from well-to-do families with successful, highly intelligent, well-off parents blessed with overachieving offspring. I, on the other hand, was a bit of a misfit, coming from a family unable to boast any of those things. In some people's minds I must have somehow winged it to get to a place like Stanford. It wasn't until I discovered a rich family history of my own that I was finally able to put some of those demons to bed.

Not understanding my place in the world was one of the reasons I threw myself into it all those years ago, leaving good jobs with great prospects, over and over again, to head off to Africa where I lived and worked for extended periods. I had an insatiable appetite to learn by doing, not by reading or studying. In a sense I was running away, if I'm honest. Looking back, I can't believe how brave I was. But being away let me learn on my own terms, an approach that gave me a degree of satisfaction and independence which proved that I had it in me, and that I was able to move far out of my comfort zone. People might be able to challenge my opinions or wisdom, I thought, but they could never challenge my experiences. The more of those I could arm myself with, the better.

I was also someone who lived very much in the moment. To me, it never mattered how well I might have done something in the past. It was all about the present – conference talks, TV interviews, writing, anything like that. I took no comfort in the fact that I may have performed perfectly well – even over-performed – on previous occasions. I always felt that my record stood for nothing. Every event felt like the first time. And every time there was pressure to deliver. I hated that.

Getting out into the world as I did paid off handsomely though, and over the course of a few years I learnt a lot about the challenges of poverty by spending time with those unfortunate enough to live with it. I found out, and experienced, far more than I ever could have by reading books on the subject. And for so much of the time it tore me up, being intimately aware of the kind of suffering happening in the world but being hopelessly unable to do anything about it.

As I was discovering, it didn't take much to feel overwhelmed with the world's problems, and once you realise the role that politics and other 'special interests' play in creating and exacerbating that poverty and inequality, it becomes even more frustrating. The more I learnt about how the world worked the less my naive belief that everyone wants what's best for everyone else stacked up. Far too many people in the world struggle day-to-day and, although many others care enough to try and do something about it, too many do nothing or, at worst, actively work to block change. One of the biggest disappointments of my life has been the realisation that not everyone wants what's best for everyone else. That saddens, angers and upsets me.

To make matters worse, many of the people I met in my work had only read books about poverty. Some hadn't even bothered with that. What I lacked in theoretical understanding I made up for with first-hand experiences, experiences which became my comfort zone, an area where all the smarter people would struggle to match me.

And, perhaps more crucially, those experiences gave me an understanding of, and credibility among, the very people I would eventually seek to help. The three years I would later spend at Sussex University getting my degree served no purpose other than to give me some academic credibility to go with it. That was how the world worked, I thought, so I reluctantly decided I should go along with it.

It was this drive to push on out into the world in search of answers and understanding that led me to Calabar, a city in southern Nigeria, in late 2001. I'd found a vacancy for a volunteer primate sanctuary manager online a few weeks earlier and met the trustee of the charity for coffee and a chat in the concourse area of Brighton train station. I had no job, was sleeping on a friend's couch and had nothing keeping me in Brighton. So, I upped and left, arriving in Calabar just over a month later.

My work there is touched on later, but a near-fatal motorcycle accident late one night on a busy highway resulted in a badly broken leg and a sudden end to my Nigerian adventure. Although I didn't realise it at the time, it also signalled the end of my lengthy search for purpose. Serendipity struck that night and an exciting new opportunity unexpectedly opened up, courtesy of that broken leg, eight days of pain, an emergency evacuation home and a chance phone call to my hospital bed.

Not surprisingly, my self-doubt and insecurity almost cost me that big chance, too. It was only two weeks before I was due to start the dream job I'd been offered during that call, and I was having one of my regular evenings up at my mum's house. She was a brilliant cook, and I remember her food was as comforting that night as it had always been. But darker clouds were looming, and the conversation quickly turned to my work, a £750,000 Vodafone-funded project I had been asked to help manage by a Cambridge-based international conservation organisation. They were looking to do something no one had done before, way back in the very early days as mobile phones were beginning to appear in many developing regions of the world. It was a time for leaders, not followers, and my mum knew me well. She knew instinctively that I was unsure of the big step I was about to take. We sat having coffee.

'So, how do you feel about moving back to the UK, Ken? It's been nice having you back here.'

'I'm excited', I replied. 'But I must admit I'm a little nervous taking on such a big job with responsibility for such a large amount of money. I've never worked with mobile phones like this before – few people have, really – so maybe I'll just end up letting everyone down.' There I went again, thinking about everyone else before me.

'Maybe you shouldn't take it then, if you're that worried.'

I knew she had my best interests at heart, but also knew she was privately hoping I didn't leave again. We'd last said goodbye like this six years earlier when I'd sold up to go to university.

I had my doubts, but I couldn't let all the pain and frustration I'd put myself through come to nothing. This was the big opportunity I had been desperately waiting for since my first tentative trip to Zambia almost two decades earlier, and Live Aid before that. I decided that night that I just had to go, to take a giant leap of faith and ignore all those negative thoughts, and that if I didn't I'd probably never be given such a great chance again. The world was telling me something, pointing me in a certain direction, and I needed to trust it, even if I didn't trust myself.

Of course, my doubts remained despite the decision to throw myself at the new opportunity. The fact that it had taken a purely random act to sort out my life, and not anything I had planned or had any control over, did nothing for my naive belief that I somehow determined my own destiny. Maybe life is mapped out for us after all, I remember thinking to myself, and we just have to go along with it – the expected and the unexpected, the good and the bad.

Sadly, many of my early life experiences fell into the 'bad'. I experienced bullying at school, and the insecurities of losing my father when I was five were compounded by the challenges we inevitably faced as a one-parent family. I often had to wear second-hand or hand-me-down clothes that would never quite fit. Despite my mother doing an incredible job under the circumstances, most days were a struggle. The teachers provided little support or comfort. I was apparently 'too sensitive' according to the school reports they wrote, as if that was a bad thing. Looking back, perhaps it was. My classes at school were dominated by children from broken or dysfunctional homes, who regularly got into trouble. It wasn't a place to be weak. Somewhere along the way I developed a stammer, knocking any remaining confidence I may have had for six. The fact I was later to stand on stages around the world and speak to hundreds of people is not lost on me. I wonder how on earth I managed any of that, too.

The result of all of this constant searching, insecurity and lingering self-doubt now sits in your hands in the shape of this much-overdue book. For reasons that are probably obvious by now, I never did relish the idea of putting down my own story, or inner workings, on paper. It all felt such a mess. Over the years I resisted doing so more times than I care to remember, despite plenty of encouragement from others who felt I really needed to share it. For reasons I'm only now beginning to understand, I've never been comfortable making anything about 'me'. I always felt more comfortable being the messenger, the enabler, the person behind the scenes. Boasting about anything 'I' might have done went against all my best instincts. Boasting is something I've always tried to avoid, believing that a lot more gets done if nobody cares about who gets the credit. The spotlight was never meant to be on me. That's obviously hard to avoid in a book like this.

So, writing a book about myself was the last thing I ever imagined doing. I assumed it presumptuous to think that anyone might be remotely interested in what I had to say, or what I'd done, or what signs there may have been in a challenging upbringing that signalled some of the successes to come. Even my editor at Kogan Page had to insist on me sharing at least some of my story as part of the publishing deal for my second book. So, reluctantly, I did. But there was so much left untold, so much history either cast aside or unknown, and so much I didn't feel ready to tell at the time.

What changed? Well, there was a whole side of me I knew so little about, even those few short years ago. I found out that I had some amazing relatives, and that their stories somehow weaved into mine. Learning about them, and becoming a parent for the first time, made the puzzle feel a lot more complete. And let's not forget the global pandemic, an event that gave me plenty of time to reflect quietly. While many people I know were spending their time looking forward, I spent most of mine looking back. I've never understood that part of me, either.

Discovering a larger group of people to write about suddenly opened up the possibility of sharing more about my journey without making the whole book about me. My children are also getting older, and they're starting to probe more into 'what daddy does for a living'. I hope this book inspires them to be as curious, brave, resilient and adventurous as I was, and that it gives them answers to questions I won't always be around to answer. I hope it also shows them that anything is possible and that humility, empathy and sensitivity are strengths and not weaknesses.

When my mother passed away in 2011, I was not only deeply upset by her death but saddened by the loss of her story, her history, her journey. What I would give for her to have written all of that down.

History will not be repeated. Sitting in your hands today is my story, my history and my journey. Putting all of this together has been a big deal for me. I hope that, in doing so, I may somehow help you, the reader, understand the ins and outs of your own journey, and how you might find your own passion, purpose and happiness in life – despite everything it may throw at you.

Part I

Purpose Pursued

Ordinary

Every story has a beginning. This is mine.

It was all pretty ordinary at the start. We were just your average sort of family. Our lives were nothing to write home about, and my life wasn't that different from the lives of many of my friends. It was neither amazing nor terrible, happy nor sad, good nor bad. It was just, well, normal. It was what it was and, knowing no different, I never really had any complaints.

I was born in St Helier on the island of Jersey in August 1966. A month earlier England had beaten West Germany in the World Cup Final at Wembley, and the country was still in celebratory mood. My dad was Welsh, so I'm not sure how he took it, but anything for a drink, I suppose. My brother had arrived less than a year before me during a prolific spell for my mother. Within the space of four years she had four children. Two more were to come after me, both girls. She later said, 'You were the best thing ever to have happened to me', but I doubt she felt that way at the time.

We lived in a state of flux, our dad often changing jobs. He was a bit of a chancer and we lived day-to-day. As a result, we had neither money nor security. My mother looked after us and did her best to keep things relatively safe, stable and sane. We'd already had a couple of homes by the time I was three – moves that we didn't make by choice – and eventually we ended up on a pretty grim housing estate with lots of other families who all seemed to be in similar positions. We didn't have much by way of toys, so we spent most of our time playing in the street outside. Most kids did, so there were regular fights and brush-ups between gangs. Bikes got stolen, things got smashed, police got called.

Our house was nothing special. The single-glazed, steel-framed windows were freezing in the winter. A couple of small, electric bar heaters were somehow meant to heat the whole place. My brother and I shared a small bedroom, and my sisters another. Bunk beds made best use of the limited space available. It wasn't a big house, and we always seemed to be on top of one another. Arguments and squabbles were common. Our numbers dropped in April 1971 when my dad suddenly died. Surprisingly, little changed after that. Clearly ordinary is ordinary, and challenging is challenging, regardless of how many parents you have.

Five years later we almost lost our mother, too. She was in hospital for what felt like an eternity. I missed her terribly, writing her letters every day begging her to come home. I was only ten and didn't understand how sick she really was. I have no idea what would have happened to us if she had died. We had no family anywhere on the island. I had friends in children's homes, so we might have ended up in one ten minutes down the road. Horrific stories of child abuse would later emerge from there, stories that would make the national news. We had such a very close escape, all four of us.

When our dad went, so did the car. He drove a beaten-up, dark grey Morris Minor, but I don't recall any of us ever going anywhere in it. I was once allowed to sit on his lap, turning the steering wheel as we inched our way around the estate, but that's about it. My mum wouldn't learn to drive for another ten years, so most of the time we were confined to the local area. Once or twice she'd muster up the courage and the five of us would take the bus to town, or perhaps the beach. But that didn't happen often, and we could rarely afford the fare. Holidays were generally non-existent. I only remember going away once as a child, on a Rotary Club day trip for disadvantaged kids to Guernsey. I remember crying with joy when I got back home.

As time went by the estate lost what little colour it had, and it became a little rougher around the edges. Small trees withered and eventually died, later dug up and never replaced. Open spaces got redeveloped to make way for new homes. It was never an amazing place, and time wasn't kind to it. The only saving grace turned out to be a youth club down the road, something which helped keep many of the children on the straight and narrow.

Nobody did particularly well at school – I, for one, had an awful time – and only one of my friends made it to university. Most of us just focused on getting by. Ambition wasn't something any of us spoke of and that didn't change much, even as we got older. Most of us got jobs, some better than others, and a few managed to escape and set up home in other parts of the island. Many didn't, though, and life took a predictable turn as they juggled bringing up kids with holding down jobs they didn't really want or like, and the lure of the pub up the road. Some years I'd go back at Christmas, only to find the same people drinking in the same seats they had sat in years before.

Culturally, I was easily influenced growing up. A late 1970s revival in the 'mod' era coincided with my early teens and I harboured dreams of being like Suggs, the lead singer of *Madness*. I'd buy cheap jackets from second-hand clothes shops and wear fingerless gloves and a flat cap, and carry my stereo around on my shoulders. Many nights I'd sit by the garages with my friends, playing *The Jam, Madness, The Specials* and *The Beat*. But I was never cool enough to be a real mod. I was far too skinny and looked silly in the clothes, and I knew it. But there was something inside me desperate to belong. My brother, on the other hand, looked the part and, for a while, he seemed to be the most popular kid on the island.

There were regular flare-ups between the mods and the rockers, so I rarely wore my jacket to town in case it got me into trouble. I had an intense fear of violence. My mum did famously break up a fight once, though, when a large, bearded, heavily tattooed rocker started picking on my brother. She called him a bully, and he seemed taken aback. He was soon gone. No one messed with our mum.

So, there you have it. Life wasn't great, but it could have been worse. I'll never know how I managed to break free from a way of life to which many of my friends resigned themselves. My sensitivity and heightened awareness were probably assets as I got older, but where I grew up they were weaknesses and certainly not strengths. Luckily I became best friends with Mark, a tall, tough, lanky, roughly-shaven character who became something of a leader to us all. I'm sure that friendship buffered me in some way from the trouble many others got themselves into.

Mark was to die young, and I broke down talking about our friendship at his funeral some years later. In all, three good friends from that estate left this world far too soon, but most are still around. It turned out that the only chance of escape would be to grab every single opportunity that came my way, an approach I was to pursue relentlessly as I got older.

So, you see, that was how it all began for me. I'd love to say there were early signs of resilience, ambition or successes to come, but there weren't. There was no reason to believe that life would turn out any differently to how it all started.

Ordinary.

Club

The skies darkened as I approached the front of our old house, my view partly obscured by the drizzling rain, a rusting white van, bikes discarded by kids and a couple of overflowing bins. Long-gone was the front lawn I used to mow meticulously as a child. There were no signs of the small rockery or the brightly-coloured pansies that once sat proudly in the centre, either. If it weren't for the number nailed along the top of the door, I'd have hardly recognised it.

Over the years, I'd talked many times about making a trip back here – a pilgrimage of sorts – but I could never quite pluck up the courage. The house and the estate it sat on held mixed memories for me, very few of them good – except for Club, that is.

Although I would have been too young to remember, everything around me was new once, untouched by the ravages of time and the continual comings and goings of young families. That's how it was in the early 1970s when we moved into number 73, a three-bedroom house at the end of a row of three. Our family had moved a couple of times after my father lost his job as manager of the Bond Hotel. A friend had stolen money while covering for my father during a bout of illness, or so the story went. The estate I was now standing on, Five Oaks, was to be our final stop, the estate where my siblings and I were to grow up, finish school, get jobs and finally leave home, and where both my father and then, many years later, my mother, would die.

As I wandered the streets that grey and chilly November afternoon, it was clear that time had not been kind to this place. The houses had never looked great, if I'm honest, all one-hundred-plus of them built to the same, uninspired architect's design, but the wooden cladding, once white and vibrant, was now weathered, drawn, peeling and coming apart at the edges. It's hard to put my finger on it, but many of the things that were once good about the place seemed to have long gone, while much of what was bad remained. It's not difficult to imagine why, after living there for over 20 years, I ended up in such a hurry to leave.

The well-sized gardens, now entirely blocked off from one another courtesy of eight-foot-high, plain wooden fencing, used to be open and inviting, a place for neighbours to glance, smile, nod and chat to one another over low garden walls. I wondered whether people weren't interested in getting to know their neighbours anymore, or whether someone else had decided it was probably better that they didn't. Small trees that used to be strategically dotted around many of the open spaces had now been reduced to tiny patches of barren earth, and the grassy areas that broke up the endless brick and concrete – places where, as kids, we regularly played football, cricket and rounders – were now people's homes, squeezed into every conceivable gap to help meet the insatiable demand for ever more social housing.

Not even the alleyways running between the houses were spared, iron gates now blocking the way, no doubt a desperate measure designed to curb antisocial behaviour. The garages in the centre of the estate, for many years a favoured spot for congregating and playing music, had been removed, leaving an area once short of character now devoid of pretty much any. I imagined us all sitting there, perched up against the aluminium garage doors, playing *Madness* out of my cheap tape recorder. Over time, everything seemed to have turned a shade of grey, from the faded tarmac and blocks of concrete to the choice of graphite-coloured bricks. It was a colour that accurately reflected the mood.

One exception to all this was the Five Oaks pub. Once a typical working-class drinking establishment (and a place where my father spent way too much time and money in the last years of his life), the Five Oaks was now an 'inn' and had transformed into an up-market gastropub. Gentrification was alive and well, even here, with the burger and chips and cheap lager long replaced by sweet potato fries, chicken Caesar salad and craft beer. We'd never have been allowed within a mile of the place if it had been anything like this in our day.

Our surroundings may have been far from inspiring, but opportunities for exploration were never far away. If nothing else, we were a very resourceful bunch. In those days, if you headed towards the far back corner of the estate you'd find an area fondly known as 'The Dump', a large patch of rough grass peppered with bushes and mounds of earth with big, highly climbable trees hugging its edges. Beyond one hedgerow you'd find a large field, a hide-and-seek paradise in the summer when it morphed into a meadow of six-foot-tall wild grass, and behind another was a narrow lane which ran alongside another large field, this one owned by a Mr Cooper (more on him later). After this was 'The Forest', a small stretch of densely packed trees with a small stream running to the side, which led into open countryside. In an era before cable television, games consoles and mobile phones, children spent most of their days outdoors in places like these, and my friends and I were no exception. Fortunately for us, outside the concrete boundaries of the estate there were plenty of ways to pass the time.

For the slightly older children, at least those lucky enough to work their way through a rather lengthy waiting list, there was Club. For an estate crawling with two to three hundred children there were surprisingly few organised social events or activities, and certainly no indoor spaces for us to hang out. Club was a saviour, for about forty of us, at least. But that wasn't the only reason it was so popular.

Frederick Richard Vivian Howard Cooper owned, ran, managed and self-funded Club, making him something of a hero among the children and parents on the estate. More fondly known as Freddie Cooper (or simply Freddie, or Mr C), his six-foot, bespectacled, well-rounded frame stood out all the more thanks to his insistence on wearing the same blue jumper, navy blue trousers and sensible walking shoes every day – his 'Club uniform', as he used to fondly describe it. He was a bear of a man and, wherever he went, feelings of comfort, safety and warmth were never far away. He was the stable father figure many of the children never had, me included.

Despite his outgoing nature, Freddie was an intensely private man, and very little was known about him despite the consistently probing questions of young, curious, nosey minds. He was Club, and Club was him, and that's pretty much all you needed to know. It was all he ever wanted you to know, too. By the time most of us left, we'd figured out a few things, mind you – that he was single, was an only child, loved Leicester City Football Club, loved working with children, was incredibly humble, kind and patient and had a wonderful sense of humour. Casting my mind back, I'll never forget his Tuck Shop 'specials', delicacies such as 'Dog-in-a-Mug' (a hot dog sausage in a plastic cup), or the way he would politely announce it was 'megazaparooney time' when Club was about to close and it was time to switch everything off. It was clear that Club meant as much to him as it did to us, and that's probably the reason it felt so special to everyone fortunate enough to come into any kind of contact with it.

Take a short ten-minute walk up the narrow, winding, tree-lined lane which ran parallel to The Dump, and you'd find yourself outside the imposing gates of St Michael's School, the largest private preparatory school on the island. We never quite figured out how Mr Cooper ended up owning the building he used for Club, which sat attached to the side of this impressive granite-built boarding school, but we were more than grateful that fate had decided he should. When Club closed years later, the school finally managed to buy it from him, but for well over a decade children from a social housing estate played metres away from kids from some of the richest families on the island, many of whom were destined to end up at private schools in the UK. We felt incredibly fortunate that no amount of money seemed to tempt Mr Cooper into a sale, and we'd heard there were plenty of offers over the years. Club really meant that much to him, and most likely gave him a strong sense of purpose that he knew money could never buy. He knew how special it was to us, too.

Club was predominantly a place of physical activity and play, and a welcome escape from our reality, with table tennis tables, skateboarding ramps, Atari video games consoles, puzzles, crafts, a set of football goals on the top field, early VHS video cassette players for watching films, cricket equipment and even full-size snooker tables. Even when Club was full, which it rarely wasn't, there was plenty for everyone to do, all supervised with a very light touch by Freddie himself. No one dared misbehave out of fear of being kicked out. It was a fun, safe, welcoming space as a result. As a former Club-goer put it at Freddie's funeral many years later, the way he ran Club taught us all how to collaborate, work together and be patient and tolerant with one another, all skills Freddie knew would be useful to us in later life.

Once the lights went out over the snooker table, and the corridors and video games machines fell silent, Club transformed into something quite different – The Learning Centre. Mr Cooper not only ran a club but was also a qualified teacher, one who specialised in helping children with learning difficulties. Most afternoons and evenings during the week he would privately tutor children who were finding it hard to keep up at school, or who needed extra help here and there. From what we could tell, this was Mr Cooper's main job and his only source of income. Fortunately for him, along with Club, it was something he not only loved but clearly excelled at.

One of Mr Cooper's secret weapons sat on a sturdy, official-looking desk in a room with huge arched windows adjacent to his main office, and this was probably one of the reasons so many parents were happy to pay for their children to be taught there. That secret weapon was a Commodore PET computer, one of the earliest personal computers (often known as 'home computers' back then). Time on the PET was very popular among his students. Unlikely as it may have been, Mr Cooper was an early proponent, and something of a visionary, in the field of computer-aided learning at a time when information technology wasn't even a subject in schools. As he reminded me many times later, he may not have understood how his computer worked, but he certainly knew how to put it to good use. To be fair, few people knew how computers worked back then, or how useful and popular they would become.

Commodore released their first machine, the PET 2001, in 1977, a year after Steve Jobs and Steve Wozniak cobbled together, by hand, Apple's first computer, the Apple I. Back then computing was more for hobbyists, with machines coming in kit form. The Apple II followed a year later, and its release, along with that of the PET 2001, opened up a new era of personal computing. Machines now came fully built, taking them out of the exclusive realm of geeks and hobbyists and into the everyday home, albeit at a price and assuming, of course, that you had some idea of what you were going to do with one.

Around the same time, Bill Gates and Paul Allen founded Microsoft. Believing the real opportunity lay in software (hence the 'soft' in its name), they went about developing programs designed to help people do things with the hardware they were beginning to buy. Interestingly IBM, who were to become the leading manufacturer of personal computers, believed the opposite, and their decision to focus on selling hardware, which they had been doing in the mainframe world for decades, gave Gates and Allen their big break. In 1981, IBM decided against developing their own software and licensed MS-DOS (Microsoft's Disk Operating System) to run on their brand new range of IBM personal computers. Within the space of a few months Microsoft had licensed MS-DOS to dozens of other companies, too, and their dominance in the operating system market later led to the development of Microsoft Windows, a platform that was to power a staggering 90% of personal computers at its peak.

In the early days, Microsoft had also been making a name for itself developing and licensing its own version of a programming language called BASIC (Beginners' All-Purpose Symbolic Instruction Code). BASIC originally emerged in the 1960s to help non-scientists and non-mathematicians program computers, a task which, up until then, required specialist knowledge. Its simplicity was key to its growth in popularity, making BASIC one of the languages of choice for many computer manufacturers who bundled it with their machines in increasing numbers. Commodore was one of these companies, choosing to license Microsoft BASIC in 1977. The PET computer sitting on Mr Cooper's desk at Club was one of them.

Commodore PETs were large, solid, heavy machines with a built-in (or more precisely, built-on) screen. The keyboard was typical of the time with big, chunky keys with plenty of travel, designed to take all the punishment they were likely to get. A cable from the back led to a small, black tape player which allowed you to load and save your code onto regular music cassettes or, if you were fortunate, a large dual floppy disk unit where you could, instead, make use of 5.25" floppy disks. You could buy a PET in any colour, as long as it was white, and any screen colour, as long as it was green.

Firing up a PET was a little uninspiring, to say the least. Within a few seconds the 40-column display would light up, and the green text would tell you how much memory was available (by today's standards, virtually nothing) and that it was 'OK' and ready to go. A square cursor blinked patiently while the computer waited to be told what to do. No mouse, no clicking, no user interface, no Windows, nothing. To make this chunk of metal do something you needed to type in a command, and the ones most commonly used were handwritten on a piece of paper stuck to the wall with oversized lumps of Blutac. Typing on the keyboard the command PRINT 'Hello World!' followed by a strike of the Enter key would, unsurprisingly, display a line of text saying 'Hello World!' followed by 'OK' and a flashing cursor below. It might seem dull now, but this was the first computer any of us had ever seen, let alone touched or used, and despite not really knowing anything about it or what to even do with it, we found it captivating.

Some of the older, more responsible children at Club were allowed to play on the PET, an incredibly brave and trusting move by Mr Cooper given how expensive it was, and how crucial it was to his teaching. That said, it was built like a tank so he probably realised it would take something special to break it. There were a few games available, if you can call them that, but none which might resemble anything you'd see today. There were no graphics or sound (other than a flat beep), either. Aliens were made out of curly brackets and dollar signs, and lasers were simple colons that moved up the screen one line at a time. Things were incredibly simple and uncomplicated, making anything seem possible.

Games could be played in a couple of different ways. By far the easiest was to grab the cassette tape for the game you wanted, insert it in the cassette player attached to the PET, and use the LOAD command, via the keyboard, to load it. Once loading was complete (which could take anything up to five minutes) and you got the 'OK' and flashing cursor back, typing in RUN started the game. This was the method that the vast majority of kids at Club used to do things.

Not surprisingly, I wasn't like the other kids. My brother and I had already been dabbling in electronics for some time, buying old TVs and radios from jumble sales and taking them apart, repurposing speakers and whatever else we could salvage. Anything left was turned into Star Trek-style control panels which we'd play with for hours as we flew through space, battling enemy ships or landing on distant planets. I was naturally curious, and the PET gave me yet another world to explore.

Very early on I wondered what other commands could be typed in after a game had loaded. After randomly coming across the LIST command in a computer magazine, and trying it for the first time after I'd loaded a game, a whole new world suddenly opened up right in front of my eyes. Laying bare were the commands that told the computer how to set up the screen and run the game and, over time, I figured out what each line of code did and how the program branched off, based on keys pressed or events that were triggered (such as dying or achieving a high score). I broke plenty of programs dabbling with the code, but it helped me learn a lot about error messages and bug fixing as I went. Eventually I was able to change things around effortlessly, make some things easier and some things harder, run the game, see where I went wrong (if I had) and try again. Mr Cooper had a simple dot matrix printer attached to the PET, and he allowed me to print off my code and take it home where I studied it enthusiastically, often late into the night under torchlight. While other children used their 'PET time' to play games, I was using mine to figure out how everything worked. It wasn't long before I started writing my own programs, and Mr Cooper's teaching turned out to be the perfect outlet for my emerging programming talents.

Most of his students required help with their reading and comprehension, and Mr Cooper supplemented more traditional paper and workbook-based activities with sessions on the PET. Here he made use of a number of fledgling education games which he bought on tape from the UK, but the students very quickly worked through them and, on top of that, they weren't always that well tailored to the specific needs of each pupil. It was around this time that I began experimenting with similar educational game ideas, putting together simple programs and sharing them with Mr Cooper.

'You know, you've got a real talent for computers', he said to me one evening as I experimented with yet another idea.

'You think so?' I replied. Everything was all so new, and I didn't often get that sort of praise.

'So, Ken, I have an idea.'

'Sure, Mr C. What's that?'

'I've got some passages of text I've been working on, just here. Could you write a program for me, with the highlighted words removed? I want to see if the students are able to figure out which words are missing.'

'I can do that, no problem', I replied, nodding confidently. The thought of someone actually using something I'd written was incredibly exciting.

Once we were confident we had something useful, he would try it out with his students who, without exception, responded positively to the new challenges we put in front of them. After a while, Mr Cooper also began providing me with sets of words which we'd break down into individual letters for the students to spell out in their minds, activities which, today, are commonplace but back then were far from it. The programs were written in a way that made it easy for me to create new ones quickly on demand and, for the first time, Mr Cooper began supporting his students with genuine, individually-tailored computer-aided learning. Because of the work I was putting in, he soon started paying me a couple of pounds per program. By now I was 14 years old and this turned out to be my second taste of financial independence. My first had been a newspaper round, delivering papers to an eye-watering 350 houses six days per week which netted me £10, a huge amount of money in those days. At one

24

point, Mr Cooper took a few samples of my work to researchers at Reading University who had shown interest in what we were doing. Unfortunately they did not take up the opportunity to work with us and I wonder, to this day, how things might have turned out if they had. These were, after all, the pioneering days of personal computing, and there was a lot up for grabs.

As my programming skills developed we expanded our suite of teaching programs at Club. At home I got hold of an early Amstrad computer and wrote a system for my mother to record her wild plant and butterfly sightings (she was a keen amateur naturalist), and started dabbling in games and databases. My interest continued to grow and, a year or so later, my skills came to the attention of the IT teacher at school who, in his other job, worked for a local computer company. I'd just turned 16, and they offered me a full-time job as a trainee COBOL (Common Business Oriented Language) programmer. Unlike Bill Gates, who famously dropped out of Harvard after just two years to pursue his ambitions with Microsoft, I decided to turn it down and finish school. Not for the first time, looking back, I wonder if I made the right decision.

By the time Club and I parted ways, Mr Cooper had written me a reference extolling the virtues of my programming skills. He signed off by saying, 'While he is still relatively young, my view would be that he could go far in having a successful and rewarding career in computer programming'.

Although I never became a professional computer programmer, computers and IT did become the cornerstone of much of my later work. My exposure to computers at Club was significant and would lead to a career operating mainframe computer systems at local banks, and the development of full-blown systems for the likes of Jersey Zoo, various legal firms and accountants and, much later, a messaging system that would impact millions of people around the world.

I did get the chance to thank Freddie, over coffee, one chilly Saturday morning in 2006 during one of my rare trips back to the island. Despite leaving Jersey ten years earlier, I had somehow managed to keep in touch.

'Well, it's lovely to see you again after all these years, Ken', he said, looking over the top of his glasses.

'I have such fond memories of Club', I replied. 'Without you, Freddie, I wouldn't be where I am today.'

He took a sip from his hot coffee, his glasses lightly steaming up, and gently shook his head. 'Oh, all I ever did was provide a place for people to go.'

'You did a lot more than that, let me tell you.'

Freddie's reluctance to acknowledge his part in my unlikely journey from Five Oaks estate all the way to Stanford University, where I was headed on a Fellowship, was no real surprise. Shrugging off any influence was Freddie all over. Whether he realised it or not, he gave me the kind of life-changing opportunity that children from Five Oaks simply didn't get. Without him, and the encouragement he gave me, my life would have been so very different.

We stepped out of the cafe and fondly said our goodbyes in the cold, autumn air. I stood there for a moment, watching him as he shuffled off down the street, my mind filled with memories of the cheerful way he had always mentored, supported and encouraged me at Club all those years ago. I could still picture him in his blue jumper, navy blue trousers, and sensible shoes.

That was the last time I saw Freddie. He died four years later, aged 70.

Rocking All Over the World

For the sort of overly sensitive child that I was, the world felt like the most unforgiving of places. For reasons I'll never understand, growing up I was often emotional, crying without ever really knowing why. I remember sitting on my mother's lap one afternoon when I was about eight or nine years old, in tears, as she tried to figure out what it was that kept upsetting me. The doctors put it down to glandular fever, and I had more blood taken from me that year than I care to remember. She never did say it out loud, but my mother must have been very concerned back then. I hate to think that I added to her worries. After all, she had plenty to deal with without me making things any worse.

Over time I learnt how to disguise my sadness by playing the fool. Joking about around my friends, something I became very good at, was almost certainly a defence mechanism to avoid having to deal with reality. For many years my strategy worked a treat. I wrote poems, too – many, over the years – as another way of escaping into worlds I could build and control. I felt so very much out of place where I was but, at the same time, had no idea where else I wanted to be, or where else I *could* be. To top it all, I was a worrier who fretted over anything and everything, often without reason or justification. I carried many of these traits into adulthood, too. Looking back, the first dozen-or-so years of my life couldn't have been much harder, but the difficulties I went through shaped me in ways I would only appreciate much later.

I did eventually realise what I was missing, and that was something to channel all of my emotion, empathy, energy and passion into other than tears. I was in need of some sort of purpose, although it wasn't really called that back then. Getting upset wasn't really the problem, but my lack of control over what was upsetting me was. I needed to feel I could control the cause, channel it, do something positive with it, and dedicate myself to rectifying it in some way. I just didn't know how.

It took a global music event, of all things, to give me what I was missing. At precisely noon one hot Saturday afternoon in July 1985, Live Aid kicked off with *Status Quo's* aptly-named 'Rocking All Over The World', signalling the start of one of the largest and most ambitious live music concerts and global fundraising events ever held. (There has been more than a fair share of debate since about the events that day, and how helpful and appropriate they really were.) As I settled down to watch the opening of the show, little did I know how significant this day would turn out to be in my life.

Over the previous two years a famine of biblical proportions had gripped Ethiopia, the worst to hit the country in over a century. These were pre-World Wide Web days and incredibly the famine, which was estimated to have impacted seven million people and killed another one million, had been kept largely hidden from view by the Ethiopian government. Described as 'the closest thing to hell on earth' by BBC reporter Michael Buerk, it was his report and film, the first by any journalist, which drove home the severity of what was happening and spurred a massive UK and global humanitarian response.

After watching Michael Buerk's report, pop stars Bob Geldof and Midge Ure quickly mobilised two dozen fellow musicians and, in a single day in November 1984, recorded 'Do They Know It's Christmas?', a hugely successful charity single released to raise urgent funds for the famine response. It became the fastest-selling UK single (it has since been overtaken by Elton John's 1997 'Candle in the Wind', a tribute to Lady Diana), selling a million copies in its first week alone and hitting the top of the charts in 14 countries, including the UK. The summer Live Aid concert was conceived as a follow-up to the Christmas single and, at its peak, boasted a global audience of almost two billion people in over 150 countries.

And sitting uncomfortably at home on Five Oaks estate, I was one of them.

My immediate reaction that day to stories and images of poverty and famine was one of shock, horror, embarrassment and guilt. Up until then I'd been largely focused on my own little world and, I hate to admit it, I poorly understood life for other people in other places. Jersey can do that to you. These days we have little excuse for not paying attention given the rise of the World Wide Web, online news and social media. Back in the 1980s, news occasionally bubbled slowly up out of the ground. There was no such thing as 'breaking news' that you could follow. Instead, an event would often come straight out and hit you like a ton of bricks. One minute there was no famine, and the next minute there was, and a biblical one at that.

In the weeks and months that followed Live Aid, I pledged to pay more attention and seek knowledge and understanding of life for the billions of other people on the planet who were far less fortunate than me. What I learnt troubled and upset me, but my determination to find some way of helping gave me the sense of control I felt I needed, driving me on despite all the emotion and frustration it caused. I shifted from being a mere bystander to being an engaged and motivated citizen. It was a subtle change, but one that made all the difference.

I began to read more widely, too. I joined organisations that focused on global conservation and development issues, and explored ways of getting involved. I wondered, apart from being a caring, considerate, compassionate member of society, whether there was anything I could do, practically, to help alleviate some of the many injustices I was learning about in the world. At that particular time in my life, technology and banking were the only concrete skills I had to offer, and I didn't find too many charities clamouring for anything like that back then. Global development seemed to be more about dam-building, and I didn't know the first thing about building dams. I figured I might just need to hang in there and simply wait for the right opportunity to come along, however long it took.

So I did.

Years quite literally passed by and, while I continued to pay attention and wait, my career in finance and information technology gradually took off. For a while life felt, and looked, almost normal.

My first job after leaving school in 1983 was as a lowly office junior at Charterhouse Japhet, which earnt me £45 per week, half of which went to my mother to cover board. After 18 months I took my first step on the career ladder when I was promoted to manage the gold, silver and platinum settlements department. The job itself was fun, far better than filing, anyway, or being constantly sent out to fetch bacon rolls for the money dealers, and I'd often find myself locked in a safe surrounded by large amounts of bullion. For someone my age, spending my days counting out millions of dollars worth of gold, silver and platinum made for good stories down the pub, but that wasn't the stand-out memory from my time at the bank.

Late one afternoon the phone rang. 'Ken, could you and John head down to reception please? I have a customer here who's interested in buying some gold.' I remember the call and the meeting well. It's not every day that someone comes in off the street looking to buy a load of gold bars, after all. It wasn't known to us, or anyone else at the time, but the man we found ourselves standing in front of was implicated in the Brinks Mat gold robbery, which had taken place a year earlier. Kenneth Noye's trip to Jersey was part of his ongoing efforts to shift the gold.

One of the largest heists of its kind, over £26 million in diamonds, gold and cash, worth over £100 million in today's money, was stolen from a warehouse at Heathrow airport, and the certificate of purchase Noye obtained from us that day would be used as cover for the stolen gold, which would be sold off in small batches. On the night of the robbery, the gang hoped to steal around £3 million in cash. They weren't expecting to find close to 7,000 gold bars as well.

Figuring out how to shift all that bullion was a challenge they willingly took on, despite neither them, nor any of their underworld contacts, having a clue what to do with so much gold. They did a pretty good job as it happens, managing to melt down and recast about half of it, most of which remains undetected in circulation today. The rest has never been recovered. As for Kenneth Noye, he was arrested following the killing of a policeman in his garden but was acquitted on grounds of self-defence at trial in 1985. The police continued to keep an eye on him, though, and a year later he was eventually found guilty of handling some of the stolen gold. He ended up serving eight years in prison, and was eventually released on parole in 2019.

After two years at Charterhouse Japhet I felt ready for a change. Next stop was a job at Hambros Bank, as a dividend clerk, where I was responsible for distributing company dividends to shareholders who held investment accounts. It didn't take long for me to realise I'd made a huge mistake. The role was boring, uninspiring and repetitive. Given my interest in computing, I looked for an early exit and saw one in the IT department. I asked James, the Director of Finance, if they needed any help and, by a complete stroke of luck, they did. They were about to start looking for someone to provide holiday cover for the full-time operator, and I jumped at the chance to do something that genuinely interested me, even if I had never been anywhere near a mainframe computer before.

The Burroughs B1900 I was about to get my hands on was entombed in a secure, glass, air-conditioned room on the second floor of the building. Halon canisters sat tucked away in one corner, ready to unload their deadly gas the minute fire broke out. Water sprinklers would have wreaked havoc on all the sensitive equipment in the room, and likely caused more damage than any fire they were supposed to put out. The B1900 itself was a beast of a machine. It was large, for a start – about a metre high, just short of a metre deep, and about four metres wide – with a small, green monitor and keyboard positioned at one end. Next to the screen was an old, cream-coloured rotary telephone (not so old back then) and a thermometer to monitor the operating temperature in the room. Along the back wall sat two large, waist-high printers. The B1900 felt alive, shaking reassuringly as its four drives took turns to power up and down. These were the days when machines had souls, and the B1900 felt like a living, breathing thing that you could almost personally get to know.

Just along the corridor outside was a narrow room where about eight data entry clerks would sit, tapping away feverishly as they entered the day's transactions from slips of paper written up by various teams around the office. The B1900 would store up the transactions and then, around 5:30am the following morning, Tony, a retired navy officer who seemed addicted to early starts, would come in and hit a couple of buttons to kick off what was known as the 'end of day routine'. All of the transactions from the previous day would be processed, and account balances updated. By the time I got in around 8am it would be close to done, and I'd print off a handful of management reports, do the backups, swap some disks around, check that everything looked sensible and bring it all to a close, ready for that day's worth of data. About half an hour after I arrived the data entry clerks would show up, eager to log into their terminals and begin the process all over again. It all ran like clockwork, with little to no room for error. It was great when it all went to plan, but when it didn't the pressure was really on. Having the whole bank sitting idly by while I tried to decode error messages and pinpoint the cause of a crash was real pressure.

After just a couple of days shadowing Sally, the full-time operator, she suddenly left without notice. Nobody knew why, but rumours circulated around the office nevertheless. I went in that morning at 8am to find James standing nervously, alone inside the computer room.

'Morning, James.'

'Hi, Ken. Good morning. You got a minute?'

'Sure. Is Sally about?'

'So, Sally has left with immediate effect I'm afraid. I can't go into the details. I know you've not been doing this job for long, but we're going to need you to step up. If this is the job you want, then it's yours for the taking.'

I'd been learning how to operate this beast of a machine for less than an hour in total, and had just one page of scrappy, handwritten notes to guide me. But I already knew more than anyone else in the building. This really was my big chance and, given my earlier experiences with Mr Cooper's Commodore PET, I knew I had it in me to succeed. That sudden, unexpected – and fortunate – resignation signalled the official start of a professional IT career that would carry me forward for the rest of my life, assuming I could navigate those first few tricky days alone at Hambros, of course.

Over the next couple of weeks I went in extra early, closer to 7am, so that I could relax a little and take my time as I worked through my steadily improving (and growing) pages of notes. The responsibility felt huge, but this was the break I needed and I wasn't going to let it slip by without a fight. As the days and weeks went by, James' confidence in me grew and I was finally offered the role full-time. I was in dreamland. As well as taking on the operation of the B1900, I helped troubleshoot the computer terminals spread throughout the bank and supported the data entry clerks. I learnt to set up and operate Burroughs B26 workstations, which were more like the standalone personal computers IBM had launched around five years earlier. I took regular trips to Guernsey where I would work with the software company responsible for the Databank system we ran, and started learning to code in COBOL. Life was good and, after the horrors of working as a dividend clerk, I was finally doing something worth getting up in the mornings for.

Within a couple of years I got itchy feet, something which happened a little too often in my early days, and I felt in need of a new challenge. Perhaps things had just got too comfortable, too easy, and I was never good in situations like that. A few other banks on the island used the Databank financial software run by Hambros, and one or two also had B1900 mainframe computers. One of those was National Westminster Bank, whose offices were a 50-metre walk away. Experienced mainframe operators that could hit the ground running were in short supply, and getting a job there was relatively straightforward. I left Hambros Bank in November 1986 and, after a short break, joined their competitor down the road.

NatWest ran two separate systems on their two Burroughs B1900s. I was given responsibility for the Databank mainframe, and my two colleagues looked after a separate system that was written and maintained by in-house London-based programmers. I was often called on to help with testing, however, and made many trips to Kegworth in the East Midlands where the bank had a huge data centre. Flying to the UK and staying in hotels was novel and exciting for me back then, as was having daily food and drink allowances. I remember ordering double-sized prawn cocktail starters most evenings. I was travelling (if you could call it that), had a job which I found interesting and rewarding, and was on pretty decent pay, all topped off with as much prawn cocktail as I could eat. What wasn't there to like?

But it wasn't all plain sailing, though, and one afternoon a freak accident almost cost me my job.

Tony, one of my best friends at the bank, was dating someone I worked with in the computer department. Fortuitously, I was dating someone he worked with in accounts. Both sets of relationships were complicated, and 'dating' might not be the best word to describe it. He'd help me out when I got stuck figuring out what was going on with Kaye, and I'd do the same for him when he needed help with Jacquie. Things were far from easy for either of us and, one day, we decided there was nothing for it but to go to the pub for lunch and drown our sorrows. We probably stayed there for a couple of hours longer than we should, and had a couple of beers too many. When we got back to work I thought it might be sensible to lay low and hide in the back corner of the computer room, out of everyone's way, until it was time to go home.

To my horror the phone, which was also stationed at the back of the room, rang. I had no choice but to pick it up. It was one of the programmers in London, requesting I load up one of the large removable drives for him, a fairly standard and regular task.

After putting down the phone I opened the cupboard behind me, took out the disk and slid open the cabinet on the mainframe, ready to load it. I carefully placed the disk in the drive and then turned the top handle clockwise a few turns to tighten it, a process which, when complete, also released the large plastic protective cover. I gently pushed the drive door shut and pressed the button to power it up.

Thirty seconds later, after a series of horrific scraping sounds, the drive shook violently and shut down. I stood there, motionless. What on earth had just happened? What had I done wrong? So much for laying low, I thought. Mike, the IT manager, had heard the noise and rushed in. Smelling my breath and suspecting I'd had one or two drinks too many, he suggested I go home and come back in the morning. I left, not knowing what had happened and whether I'd have a job to return to.

The next day I arrived back at work and, fearing the worst, headed straight for Mike's office.

'Morning, Mike.'

'Hi, Ken. Take a seat. How are you feeling today, then?'

'I'm fine thanks, but obviously worried about what happened yesterday. I'm really sorry. What exactly *did* happen?' I really had no idea. The operation I was asked to carry out was simple, and one I'd done many times before.

'I'll explain in a minute, Ken. But first of all, can I just say that if you decide to go out for a long, liquid lunch again, you just go straight home afterwards. It's never a good idea to be around the mainframes if you've had a few too many to drink.'

I nodded. 'Sure, Mike. Sorry.'

'So, this is what happened yesterday. I don't think the beer played too much of a role, but there's a chance you may have spotted the problem if you'd been sober.' I sat, staring straight ahead, wondering what he was going to say next.

'It turns out, quite unbelievably, that there's been a leak in the external flat roof above the cupboard where we keep the drives. Rain had got through and onto the drives inside. The one you loaded had water on it, basically. And the water caused the heads to crash.'

(It might help to pause for a quick science lesson here. Whenever you power up one of these drives, the spinning motion creates a cushion of air above the platters (or disks), and once up to speed a series of heads, which read the data, slide out. The heads never come into actual contact with the drive, but the distance they float above is minuscule. The water droplets would have collided with the heads, causing them to vibrate, drop down and make contact. That was the horrific noise I'd heard, and never wished to hear again.)

I couldn't believe what Mike had just told me, and wasn't sure how to react. Should I be relieved, shocked, surprised, or a combination of all three?

Mike let me off with a warning.

I didn't break anything again after that, but within a couple of years I once again got the itch. It seemed I couldn't properly settle anywhere. Working in IT in the finance industry didn't stop being fun, challenging or rewarding, but it was restrictive. I only got to work on hardware and software approved and commissioned by the bank, and there was a whole world of personal computing beginning to take off outside. I wanted more variety, and to hone my skills on a wider range of equipment in a broader range of settings.

In October 1988 I left the relative comfort and security of the finance sector and joined a two-man tech startup. My salary halved, but it seemed like a price worth paying given the opportunity. Most of my friends thought I was mad, mind you, as did my mum. I was quickly thrown into engineering tasks I had zero knowledge of, learning on my feet to hand-solder printer T-switches and Ethernet sockets at clients' sites. I helped install and maintain Novell networks, and was frontline tech support for dozens of clients, ranging from financial institutions to Jersey Airport. I was often left on my own, and if I broke anything I'd usually be left to figure out how to fix it by myself. It was scary and exciting. You really can't beat being thrown in the deep end like that.

After just one year the firm went bust. Mark, the CEO, enjoyed fast cars and fine dining, and spent a little too much money obtaining his private pilot's licence. I wasn't paid for the last two months, and ended up unemployed. Fortunately, one or two of the customers also ended up with computer systems no one else knew how to maintain, and one of them – International Law Systems (ILS) in Guernsey – moved fast to find a way of keeping me around. I'd been heavily involved in the setting-up of their desktop publishing system (they were a specialist print and design shop for the legal profession) and Chris, the entrepreneurial CEO of this great little family business, was keen to find a role for me.

We bought back some of my old furniture and computer equipment at the bankruptcy auction, set it up in my bedroom and incorporated Clifton Computer Consultants. I took on the grand title of Managing Director and loved running my own business. I was just 22 years old at the time. When I wasn't flying back and forth to Guernsey helping ILS, I was out selling and building up the desktop publishing business. Within a few short months I secured the Channel Island dealership for Rank Xerox's Ventura Publisher desktop publishing system, much to the horror and surprise of the larger, more established computer companies on the island who also had their eyes on it. (Ventura Publisher was, at the time, competing hard with Apple Macintosh computers, the system of choice for serious designers, or so we were told.) A few months after that we moved out of my bedroom and set up shop in a small office at the back of a doctor's surgery in St Helier, and then, as things picked up further, relocated to a larger office above a clothes shop in the middle of town. To add to the Ventura dealership, I also became the Channel Island authorised dealer for AST Computers.

Clifton Computers gave me my first taste of running a business at a time when home computing was really beginning to take off. My bet to leave the safe haven of the finance sector was paying off, and I was the master of my own destiny. I managed to put my coding hat back on, too, and developed an office automation system for ILS which ran all of their sales, ordering and accounts. This experience was to prove invaluable as I later developed bespoke systems for other firms on the island, including Jersey Zoo.

One year in and the company was doing well, but the recruitment of a sales manager who ended up selling nothing put huge pressure on us financially. I wanted to replace her but Chris didn't and, given he was our biggest customer with an annual support contract that effectively kept the company afloat, I caved. Things didn't improve, though, and just over two years after we'd formed the company it was clear we needed to either invest heavily or call it a day. Back then I didn't have much of an appetite for debt – I'd previously got myself into a mess at a young age with store cards, credit cards and personal loans – and Chris decided to focus his efforts on opening new satellite offices for ILS. An approach to my grandfather for investment (he had already put money into my uncle's computer business a few years earlier) disappointingly came to nothing. In January 1992 we closed down Clifton Computers but, to maintain continuity, I kept working for ILS for another six months while I sought out my next opportunity, supporting and developing their IT systems from home.

My next stop was another privately-owned business machines company, albeit one with more of a focus on photocopiers and fax machines than computers. Vic, the owner, heard I was looking for work from a diving friend who had been a customer of mine at Clifton Computers, and saw an opportunity to move into personal computers with me on board. His hunch was right, and business boomed as we combined his healthy island-wide customer base with my hardware and software expertise. I was precisely the sort of all-rounder he needed, and we thrived. We developed accounting software for legal firms, internal mail messaging systems (about five years before email finally took off) and, without doubt my proudest achievement, a fully networked membership and management system for Jersey Zoo. I continued to work on this for a number of years, even after Vic and I had parted ways, which we eventually did in late 1994. Working late into the evenings at Les Augres Manor, home of the Zoo, with animal calls ringing out into the night, helped make the best kind of memories.

I didn't realise it at the time, but my last job on the island before selling up and leaving for university would be at JerseyCard, later to become SuperCard following a takeover by Nick Ogden, a serial entrepreneur who would go on to create WorldPay (losing it some years later in a hostile takeover). SuperCard was a small company, and our main product was a smartcard-based loyalty scheme that ran off electronic terminals which could also take regular debit and credit cards. We would program the terminals and ship them out to clients, mostly shops and restaurants, and then suck up all the transactions over the phone line for processing each night. It wasn't an amazing job, but I spent a lot of time configuring and repairing devices, and visiting customer sites, both things I really enjoyed doing. We were based in a shared office with a bunch of other people doing groundbreaking work in the world of online payments and e-commerce, which also kept things interesting. Nick also ran Supernet, one of the first Internet Service Providers (ISPs) on the island, and he also built and launched what is believed to be the first-ever online shop, the Wine Warehouse, in 1994.

Despite enjoying the work, itchy feet once again got the better of me. Live Aid was in danger of becoming a distant memory, and I needed to get back on track. Over time it had become obvious that my future lay off the island, and my next move was to be the biggest and most disruptive to date. I had no plans to return, so my final act at SuperCard, in late August 1996, was to invite everyone I knew around for a flash sale at the small, three-bed cottage I was renting in St Mary's, a beautiful rural parish out of town. With the exception of the bed I was still sleeping in, a chair and a single set of knives, forks and plates, everything I owned was up for sale and, in the space of an hour, the place got cleared.

That night I sat alone in the garden of what was once a cosy home, cold beer in hand. I reflected on the previous few years and the career I was now turning my back on. I'd made some progress, but the one thing I knew I needed more than anything was some kind of experience working in the developing world. Fortunately that had been taken care of, courtesy of a tourist train and a random event one afternoon in a local town pub. But more on that in a minute.

For now, I felt liberated at the thought that I could fit everything I owned into two average-size suitcases. All that now remained was to officially hand in my notice at work, and then the next chapter in my search for purpose could begin.

Jersey Potatoes

'I can't quite believe we're about to walk around the island', muttered Ray as he took one final gulp of his beer, carefully sliding his blue rucksack over his shoulders before helping me with mine.

'I'm not sure I do either, Ray. But I'm sure one day we'll have a good laugh about it, assuming we come out of it alive', I joked, retying my laces. Whatever happened, at least I'd get to spend some quality time with Ray. I always enjoyed his company and his way of looking at the world, and we'd have plenty of time to explore all kinds of subjects over the coming hours as we dragged ourselves around the island. It would be just me and him, and miles of open road.

Luckily for us, Jersey happens to be a great place if you're into walking. The largest of five islands in the English Channel (seven if you include Jethou and Brecqhou, which between them amount to less than a square mile), Jersey boasts incredible natural beauty, the best beaches, great weather, low unemployment, low crime, low taxes, good education and healthcare, and its own language (more on that later). Although technically part of the United Kingdom, Jersey has its own government and can do pretty-much anything it wants. As it turns out, that's not always a good thing.

With a scattering of claps, cheers and a fair few looks of amazement – yes, we really were about to do this – Ray and I broke away from the group and crossed the road from the town bus station towards the seafront. Weekend party-goers sat alongside holidaymakers, eating and drinking in nearby bars, huddled tightly around tables put out on the pavement to make the most of the warm evening sun. Music pulsed from every direction. Years ago I would have been sitting among this mass of people, enjoying a typical Saturday night out, but tonight was going to be quite different.

As we waved goodbye we didn't know precisely when we'd see everyone again. As it happened, we never did quite make it all the way around the island. Probably three-quarters of the way would be a good guess. We ended up getting rescued close to 12 hours later, a taxi taking us the last few miles back to town. If we'd tried going on any further the general hospital would have been our next stop. By the time Gorey Castle was slipping out of view behind us, Ray was dead on his feet, and I wasn't too far behind.

A month or so earlier, over a few too many drinks, we'd all talked about how we might raise awareness of our trip to Uganda that autumn, and how to bring in some money to help with extra supplies. Although the trip was heavily subsidised by the Jersey government, we still wanted money to provide extra equipment to the hospital we'd be visiting, and for stationery and books for the local school. We were told something as simple as a pen and exercise book would be prized possessions for the children there, and we'd later find out for ourselves how true that would be.

It was quite some time ago, and we were all a little worse for wear, so I can't say for sure where or who the idea came from, but the potent mix of lofty ambition and cheap beer were key contributors. By the time the bell rang for last orders, Ray and I had agreed to, or at least been convinced to, walk around the coast of Jersey with six pounds of potatoes strapped to our backs. We'd have a few weeks to raise awareness of the walk, seek sponsors and buy the potatoes, and maybe get in a little training, if we decided we needed any.

In much of the developing world people walk incredible distances with all manner of things balanced on their heads, gripped in their hands or strapped to their backs, or sometimes a combination of all three. On my own travels I've seen everything from fridge freezers to coffins (empty ones, I've always hoped) rocking to and fro on people's heads as they've gone about their daily business. In our team's wisdom we decided that walking around the island, carrying the equivalent weight of a baby the whole way, might represent the hardship many African mothers endure as they carry their newborns on their backs while working the fields, going to market, or collecting water or firewood under a hot sun. We weren't sure what to expect but had a hunch, a good one, as it turned out, that it wasn't going to be easy.

Despite everything, we opted to do absolutely no training beforehand. We did do a little planning though and, knowing we'd be walking on roads for much of the way, decided to attempt the walk one Saturday night going into Sunday morning, when traffic would be either light or non-existent. This also had the effect of helping us avoid the heat of the day, something which did little to recreate the reality of the situation for all those African mothers.

Ray was our team leader. He was a short, stocky and deeply religious man. I enjoyed the way he spoke about his faith, and appreciated the fact that he didn't ever try to convert me. 'God will find you when you're ready', he always used to say. Sadly, Ray passed away a few years ago, and God still hasn't found me, in case you were wondering. This trip was the latest of many he'd taken to Soroti, a town in Eastern Uganda with a population of around 40,000 people. The States of Jersey Overseas Aid Committee had been supporting the building of a medical centre there for many years and Ray regularly led trips, with a new band of volunteers each time, to help move it along. This year I would be among them.

This wasn't my first visit to the African continent, mind you. My first ever trip, to Zambia two years earlier, transformed me. Reading about poverty – seeing it close-up on television, even – was one thing. Witnessing it, smelling it and tasting it was another thing altogether. I challenge anyone to see suffering up close and not have it affect them in some way. For someone as sensitive as me, they were images I'd never be able to forget or let go.

Although I'd been on the lookout for opportunities since Live Aid, I'd slowly drifted into a life of regular work in the finance industry, and that first Zambian opportunity came completely out of the blue. Events leading up to me learning about the trip were nothing short of miraculous and, looking back, it's easy to believe that fate must have played a big part.

It was early March 1993, and spring was in the air. It was one of those glorious, crisp, bright, sunny days that you long for after months of winter weather. It was midweek, and I'd arranged to meet my girlfriend, Kaye, at the Prince of Wales pub for lunch. These were the days long before mobile phones and, not only had I messed up the time (we were supposed to meet at 1pm), I'd also gone into the wrong pub. As I sat waiting for someone who was never going to show, the local news came on the radio. My ears pricked. Local politicians were discussing whether to allow a diesel-powered tourist train, going by the name of Terence the Train, to run on a section of cycle track that stretched from St Aubin to St Helier, sweeping and curving its way along the coast. At the time I was a keen cyclist, riding into work most days and in most weather from my home a few miles away. The thought that one of the only places I could ride safely, without worrying about cars, was about to be invaded by a tourist train incensed me.

That evening on the way home I grabbed myself a copy of the *Jersey Evening Post* to get the full story before penning a letter of objection to the newspaper's editor. I rarely bought the paper and wasn't the sort of person to air my grievances without good reason, but the fact that many other people seemed to feel the same helped make the decision for me. My letter – 'Keep Terence Off Our Cycle Track' – was published a week later and, not unsurprisingly, achieved little. Loved by the tourists and hated by the locals, the train began chugging its way up and down the cycle track in no time at all, something it continues to do to this day.

My time wasn't totally wasted, though. On page two of the paper that night was an advert from Jersey Overseas Aid promoting a forthcoming trip to Zambia. They were seeking volunteers to continue the building of teacher accommodation for a local primary school. Kevin and Vanessa, the two leaders of the trip, had been returning to Chilubula, a village in the north of the country, for a number of years as work had progressed. They needed ten volunteers to take with them on a five-week trip that July. It looked like an incredible opportunity. It wasn't impossible to imagine that I could get five weeks off work and, at around £350 (the trip was heavily subsidised) the cost was far from prohibitive. Could this be the breakthrough I'd been looking for? I had nothing to lose, so the next day I excitedly sent off my application.

I remember the interview well. It was, after all, a defining moment in my life, and failure would almost certainly have signalled the end of my ambitions. It was a warm day, and I remember walking hesitantly into the classroom where I found Kevin and Vanessa sitting behind a couple of desks, paperwork laid out in front of them. They were very easy to talk to and, despite my nerves, they later told me I came across as relaxed, easy-going and genuinely interested in the work. There were plenty of smiles, and enough laughter to give me confidence that I might stand out from the many other applicants.

'Hi, Ken. Take a seat.' I smiled and pulled up a chair. 'Thanks for your interest in the Zambian trip this summer. Maybe you can begin by explaining a little about yourself, and why you think we should pick you.'

I took a deep breath, and started by talking about my upbringing and schooling, and then how Live Aid, a few years earlier, had piqued my interest in global development.

'I'm really interested in international development, and am keen to learn more about the work', I said.

'And what do you hope to get from doing that?' asked Vanessa.

'I think it's important to understand what life is like for people in these places first and foremost. Ultimately, I hope I can make a difference, I suppose.'

'Do you think you can?'

'I'm not 100% sure at the moment, to be honest. That's one reason for wanting to join you on the trip. But I'll give it everything if you decide to take me.'

Vanessa nodded reassuringly, sliding a couple of photos across the table towards me.

'We've been visiting Chilubula, a village in the north of the country, for a few years now and this year we'll start work on living accommodation for the teachers at the local primary school, which you can see in these pictures. If you're picked you'll be helping us build them.'

Kevin leaned forward, pointing to one of the pictures. 'The thing is, many teachers are trained in Zambia, but then they leave to teach in places with better resources and higher salaries', he explained. 'The purpose of building homes for the teachers is to make staying a little more attractive.'

I must admit, I did wonder why a bunch of people from Jersey were flying all that way to lay some bricks when there must have been people in Chilubula who could do it just as easily, likely better. But Kevin's earlier comments about these trips being just as much a cultural exchange, and an opportunity for islanders to experience other parts of the world, sold me. Later, as I learnt more about how the development sector operates, I became increasingly uncomfortable with this approach, despite the obvious tension with how it all began for me.

As we neared the end of my interview I assured them that I had the money – I nearly did – and that I had permission to take the time off work. The latter wasn't strictly true. I'd decided not to ask Vic for so much time off until I knew I'd be going.

Within a couple of weeks news came through that I'd been selected. This was a defining moment, not just in my future career but also my life. I was absolutely over the moon to be getting my first taste of development, finally, after a long, eight-year wait. Not only that, for the first time I would also find myself in the company of others with the same interests and passions and, with this trip, a common goal to work towards. Over the next couple of months our team met regularly, hatched plans to raise money, and we all became good friends. Thanks to our regular meetups, often down the pub, we got very good at finding fun and ingenious ways to get other people to support the trip.

A friend of mine, Nigel, played in a covers band, and one Saturday afternoon he kindly set up his equipment on the back of an open truck, driving around and stopping at popular spots on the island to play. As he and his singer blasted out the hits, the rest of us mingled with the dancing crowds, shaking Jersey Overseas Aid collection tins. Nigel's efforts brought in hundreds of pounds, an incredible achievement that we duly celebrated that evening. I was really enjoying myself by this time, and I soon put my artistic skills to good use designing the trip t-shirts. I also became treasurer. My passion and enthusiasm for the trip, and my organisational skills, were already beginning to shine through.

The next few weeks passed quickly by as we got visas, airline tickets, clothing, supplies and vaccinations – lots of vaccinations. This was my first real overseas trip, and I needed jabs for absolutely everything before I'd be allowed on the plane. Only the rabies jab gave me any trouble, a restless, nightmare-ridden sleep from what I remember. The only thing I was really worried about was the Lariam, the tablets we'd take to protect us from malaria, which had a list of side effects as long as my arm. Most of us didn't feel great taking them, but we were all new to this so largely did what we were told.

As the morning of our departure arrived, I penned a one-page letter to my mother which I left on the coffee table in the lounge of the one-bedroom flat I was renting on the outskirts of town. I really didn't know what I was letting myself in for, and genuinely didn't know whether I was going to get back unscathed, or even get back at all. As it happens, I very nearly didn't make it home – more on that a little later – but I wanted her to know what an inspiration she had been, and how grateful I was for everything she'd done for me. I ended by telling her not to be sad if anything bad should happen to me.

We arrived in Lusaka, the Zambian capital, the following morning on an overnight flight via Heathrow after a pleasant day taking in the sights and sounds of Covent Garden. Few of us had been abroad that much, so trips to London were an event in themselves. It was a final chance to bond and relax before the real work began. None of us slept much on the journey over, a mixture of excitement and apprehension keeping most of us awake for the roughly nine-hour flight.

I'll never forget taking those very first steps from the comfort of the air-conditioned cabin into the warm, humid, stuffy Zambian air outside, the taste of dry, earthy dust hitting my lips and the bright sunshine forcing my eyes hard shut. As I reached the bottom of the aircraft steps, I placed my feet on African soil – albeit airport tarmac – for the very first time, a moment which felt overwhelming after everything I'd read about the continent, and the seeming impossibility of ever getting there. Later, I would carefully wrap a pair of white, unwashed sports socks into a sheet of newspaper as a reminder of the incredible red, clay-like soil we trod daily in Chilubula. I still have those socks, safely tucked away to this day and, disgusting as they look, they represent some of my earliest steps on a continent that would end up playing a huge part in my life.

Lusaka is the largest city in Zambia and, when we arrived in July 1993, it had a population of about a million people (it now exceeds 3 million). Livingstone, close to Victoria Falls, was the first capital city but, in 1935, it was decided that Lusaka would be more suitable given its central location and proximity to main rail and road routes. It was unlike any place I'd ever been. It was by far the hottest and most polluted – the traffic fumes were suffocating – but it was the hustle and bustle and sheer volume of people everywhere that were most overwhelming. I was struck by the number of people begging on the streets. Many sat, in various states of desperation, hands outstretched, weeping. Just walking away was one of the hardest things I had to learn to do early on. It wasn't in my nature to ignore this kind of suffering, and I would have liked to do something for everyone I passed. Sadly, and not unsurprisingly, there was little I could do given the scale of what I was witnessing.

We stayed overnight in Lusaka, and the following morning headed back to the airport to take a domestic flight to Kasama, roughly two hours away. The total flight time would have been closer to an hour had it not been for the two extra stops we had to make, stops that did at least give us the chance to see a little more of the country at an altitude where you can actually see something. The medium-sized, twin-propeller aircraft we boarded was older than many other planes I'd been on, but it seemed sturdy enough. I decided not to think too much about the last time it might have been serviced or had a safety check. Just four months earlier a plane not too dissimilar to the one we were on, carrying the Zambian national football team, crashed into the Atlantic Ocean with the loss of all thirty passengers on board. We would later watch a match on television, played by a makeshift Zambian team, which conjured up all sorts of emotions for those we watched it with.

I remember two things in particular about our journey that day. The first was the chickens that one passenger carried aboard as hand luggage, one of which escaped into the aisle during the flight. Nobody seemed bothered by this, if I'm honest. Clearly, it was nothing out of the ordinary for regular passengers on that route, or perhaps on any domestic route, to have chickens roaming the cabin. The second was the sight of a lowly airport fire engine sitting propped up on bricks near one of the terminal buildings. The first thing I wondered was whether it still worked and, if it did, how quickly they'd be able to get the wheels on if a plane did crash-land and catch fire. I didn't hold out much hope for any of the passengers if it did. I was told by a fellow passenger that the wheels had been removed to prevent them from being stolen. To this day I still don't know whether he was telling me the truth, or whether he just couldn't resist the opportunity to play a joke on a random white guy.

We were met at the airport in Kasama by an old, worse-for-wear white Toyota pick-up truck which, on this occasion, took our luggage but would later ferry the team around. With our bags safely loaded, we all got onto a minibus for the 40-minute drive to the compound in Chilubula where we'd be based for the next four weeks. The fifth week of the trip was reserved for a short break at Victoria Falls.

The drive to Chilubula gave us all a chance to see rural Zambia close up for the very first time. The outskirts of the airport resembled a forest of billboards, selling everything from soap to fizzy drinks to televisions, while others pointed to the field

46

offices of the many non-profit organisations working in the area. As we drove further and further north things began to thin out, and only small settlements broke up the endless sight of grassland and shrub. I remember being struck by the vastness of it all, and how dry and barren everything was. I also remember seeing my first grass-roofed mud hut. For a naive first-timer, that single image represented the African continent more than any other. I took photos of them, too. Looking back it seems silly, but everything was so new to me back then. I had waited an awfully long time for the opportunity to see, taste, smell and experience a place like this. It didn't disappoint.

By late afternoon we arrived at the walled, gated compound we'd call home, and were immediately greeted by ecstatic primary school-aged children who shouted, jumped, screamed and sang as our minibus gingerly pulled up inside. Strangely, amid all the noise and excitement, one child caught my eye. He was standing quietly and thoughtfully, carefully watching everything going on around him. I turned to one of the local church members who had also welcomed us, and asked who the child was. 'Oh, that's Chiti', came the reply. I'd meet Chiti, and his father, properly a little later on the trip.

We paired up and shared bedrooms in the complex, which also had a basic kitchen and dining area. We had someone come in who cooked for us, and someone else who cleaned and kept the whole place in order. This was something I had to get used to. I wasn't sure how I felt about being waited on but, as I was reminded on that trip and many after, these were much-needed employment opportunities for local people and it would be wrong to deny them jobs because of a 'little discomfort'. For me, questions of dignity and power would come up time and time again during the course of my humanitarian career.

Chilubula Primary School was only a short drive away down a winding dirt track, and we'd leave at the crack of dawn each morning, packed into the back of the truck, to go and work on the new accommodation. The school itself was modest, not unlike many you'd find in most rural villages, at least those fortunate enough to have one. There were only a couple of classrooms and, in each one, only a handful of desks and chairs, far from enough for all the children. The hard, dusty concrete floor and whitewashed walls made it feel strangely cold, despite the soaring temperatures outside. The wall at the front, next to where the teacher would stand, had a rough rectangle across the middle painted in blackboard paint. Large parts had faded and it was obvious that, over time, the chalk had become increasingly difficult to remove. There were no books in sight, neither reading books nor exercise books, and little by way of a school uniform if what the children were wearing was anything to go by.

Primary education was free in Zambia back then – it may still be – so the classes always seemed full. When costs kicked in for secondary education, though, many parents struggled to pay the fees. As a result, many children dropped out with only a very basic understanding of maths, English and other subjects. Almost all the parents I spoke to were desperate for their children to go on to secondary school, knowing that an education was the best chance they'd have for a better life. Many of the children, too, were desperate to learn. All of this sat uncomfortably with me knowing how many children back home had to be forced to go to school, and how much I, personally, had struggled to make the most of what I had, despite all the opportunities given me.

Most days passed by without incident as we mixed cement and slowly built up the walls. With about a dozen of us on site most days, and the houses only being single storey, we made rapid progress and, as we neared the end of our trip the metal-framed windows went in and the building was ready for the roof to go up. Some days I was pulled off-site to help buy building materials or to withdraw project funds from Standard Chartered in Kasama. It was nice to get a flavour of some of the management needs of a project like this, knowledge that would come in handy in the future.

When we weren't on site we'd relax, write, read or listen to music. Many of the local children would stare in awe as I loaded up my portable CD player. They'd never seen anything like it, and couldn't figure out where the music was coming from. When we had more time we'd visit local attractions such as Chishimba Falls, about a 20-minute drive away from Chilubula, or take day trips into Kasama to enjoy the market or the bars and restaurants. Occasionally, I'd be left behind at the compound as everyone else headed off to the building site. We'd learnt that the local hospital had no record keeping system so couldn't track patients being admitted or discharged. Kevin, Vanessa and I had decided back in Jersey that it was probably worth bringing my BASIC disks with me, just in case anybody needed some kind of IT help. The hospital did have an old IBM PC, which was brought over to the compound for a week while I wrote a simple system for them. When it was finished I held a short training session for the senior staff and handed it all over.

To this day, I don't know if it was ever used and, if it was, for how long. Given everything that was to happen later, it's funny to think that I undertook my first technology project in Africa, however small or unplanned it may have been, all the way back in 1993.

Given the trip was also designed to be something of a cultural exchange, each of us were offered the opportunity to spend a night alone with a family in the village. This was my first time doing anything like this and, despite being understandably nervous, I didn't want to miss the chance. What better way to get a sense of what day-to-day life was like for people, I remember thinking to myself. I stayed with one of the teachers from the local primary school, arriving for dinner that evening and leaving after lunch the following day. I didn't realise it at the time, but they'd killed one of their best chickens to provide me with the three meals I'd be sharing with them, although I use the word 'sharing' loosely. It became clear to me, as time went by, that I was the only one really eating the chicken, at least the bits that you'd consider edible. The children and the mother waited out of sight each time, while I ate with the father, something which made me very uncomfortable. Whatever was left ended up on their plates, along with the head, feet and any of the insides I might not have wanted.

The longer the trip went on, the more relaxed I felt in my surroundings. Nothing I'd witnessed up to that point put me off wanting to do more, although I had seen things I'd found challenging, things that left me with lots of questions. I remember one event that particularly troubled me. After finishing work early one day, we drove to a more remote, neighbouring village. We pulled up and got out of the truck. Out of nowhere a man appeared, running over. 'Mzungu! Mzungu!' he cried, throwing himself down on the ground in front of us. (Mzungu is a Swahili word, and it can mean a few things, but it's most commonly used to describe a 'white person'.) 'It's so wonderful to be graced with your presence here', he continued. It turned out he was the village chief. We encouraged him to stand up, which he eventually did, yet his head remained bowed for the rest of the conversation.

This single act drove home more than most the power dynamics at play between local people, who often needed a huge amount of help, and us, outsiders who were seen as having all the money and resources at our disposal to provide it. I had no intention of playing the White Saviour, but sometimes it was hard not to be seen as one, however hard I tried.

As we approached the last few days of our time in Chilubula, I started to think about what I could buy as a memento of what increasingly felt like a life-changing experience. I didn't want anything from the tourist shops, or anything produced for the tourist market. I wanted something that felt more authentic. I learned of an artist who lived close by and someone suggested I go and see him. This was the first time I met Justice Kabango, and we were to become very good friends. Chiti, the quiet boy I'd spotted on our first day, happened to be his youngest son. He was also there when I called round.

Justice had four sons and juggled jobs as a mechanic, a handyman and an artist to try and keep them all in school. His wife had died a couple of years earlier from an HIV/AIDS-related illness, and I admired how hard he now worked, and how determined he was, to provide for his family and keep everyone happy, healthy and together. Buying a painting from him felt like the perfect way to not just help him, but to get something meaningful that would remind me of my time in his country. I found a picture of a couple of lions, painted on a thin, roughly-cut piece of wood which was then glued to a thicker, larger piece, that seemed ideal. I didn't haggle over the price but paid him what he asked. Before I left we swapped addresses, and promised to keep in touch.

As our Zambian trip drew to a close we had one last stop in Livingstone before heading home. Founded in 1905, Livingstone is named after the intrepid Scottish explorer, Dr David Livingstone, the first outsider to see what was to later be known as Victoria Falls. Originally called the Old Drift and originally founded as a staging point for travel across the Zambezi River, Livingstone itself is a relatively small town. Despite its size it's often referred to as the tourist capital of Zambia, primarily because of the popularity of the Falls, the very reason we were headed there ourselves.

After we'd packed up and said our goodbyes we were driven back to the airport in Kasama. As we wandered through the departure hall, bags dragging behind us, Kevin and Vanessa checked the board for our flight to Livingstone. Nothing. They checked our tickets. We were at least a couple of hours early, as you're meant to be. Kevin told us to wait while he and Vanessa headed to the check-in desk to find out what was going on.

'We're booked on the 16.30 flight to Livingstone, but I can't see it on the board', explained Kevin. It turned out there was a good reason it wasn't there.

'Oh, the Livingstone flight left at lunchtime', the check-in assistant replied. The flight time had changed, but nobody had told us. There wasn't much point in arguing about it. This was how things rolled in Zambia, and we'd just need to deal with it.

Plan B turned out to be a coach ride to Lusaka, then a plane on to Livingstone. The change of plans, plus going overland, added to our sense of adventure and we got to see a lot more of the country as a result. It did take us a little longer than originally planned, but we arrived in Livingstone in one piece the next day where we stayed at an open hotel complex in fancy, grass-roofed huts within earshot of the Falls. We spent the rest of that first day relaxing and visiting the many tourist shops nearby. Our first full day would be taken up whitewater rafting, a day I'll never forget for all the wrong reasons.

I'd never learnt to swim at school, and would only learn much later. Our sports teachers were only really interested in working with kids who had a chance of winning something, and I wasn't one of them. Instead, the non-swimmers were largely left to our own devices, splashing around in the shallow end and not really doing much. I was afraid of water back then and had no confidence either in it, or anywhere near it. It didn't help that, occasionally, the teachers would throw us in – maybe for a laugh, who knows – and that drove my fear yet further. So the idea of going whitewater rafting on a raging river wasn't something I was overly thrilled about, as you can imagine, but I convinced myself to do it. I didn't know whether I'd ever get the chance again, and everyone else seemed up for it.

That morning we hiked down the steep rocky bank towards the edge of the Zambezi River. It was a beautiful day, made all the more special by the rolling mist, the roaring of the Falls and the spectacular backdrop of Victoria Falls Bridge spanning high above us. Victoria is one of the largest waterfalls in the world, coming in at a whopping 5,600 feet wide. Every second close to 40,000 cubic feet of water drops a full 354 feet, creating a mass of bubbling, swirling, crashing water below. With the exception of the odd calm and often incredibly scenic stretch, patches of white water crash and swirl for the next 17 miles or so down river.

We'd be tackling the first couple of dozen rapids that day, although there are a few more after that for diehard rafters. Each rapid has a number and a name, everything from Creamy White Buttocks or The Gnashing Jaws of Death, to The Terminator or The Washing Machine. Rapid number nine – the aptly named Commercial Suicide –

is the one everyone skips. Instead, rafts are dragged out onto the bank ahead of an awe-inspiring walk around it. A ferocious grade six (the highest possible), this rapid is considered way too dangerous to be commercially viable.

After some safety advice, which included strict instructions to 'hang onto the boat whatever happens', and 'in no circumstances let go of the rope running around the edge of the boat', we put on our life jackets and climbed aboard two separate rafts. The water already looked intimidating, and we hadn't gone anywhere yet. But things started well, and we navigated the first four rapids without too much trouble, throwing ourselves around the raft as we desperately fought to stay balanced, upright and afloat. It was exhilarating, a huge adrenaline rush to match no other. I tried to enjoy it but was scared to death. Everything about the situation told me that I shouldn't be there and that I was stupid to have agreed to do it. During those moments, the letter I'd left at home for my mother suddenly felt like a good idea.

My fears were realised when we hit Stairway to Heaven, the fifth rapid and our first grade five. It began like all the others, with the raft being thrown around all over the place, left then right, up then down, front then back as we hit the water at different angles and speeds in quick succession, over and over again. The crashing of waves and the swirling of water was deafening as we tried in vain to work our way around to the left. Go too far and you hit a huge wave known as the 'catcher's mitt', which is apparently similar to dropping off a two-storey building. Hit that wave we did and, with a huge thud, the boat pointed skywards, things went dark and we were suddenly upside down. Everyone managed to keep hold of the rope, a move which kept them attached to the best buoyancy aid we had – the raft.

Everyone except me, that is. My instinct told me I was in danger, and the best thing to do was to get away. Unconsciously, I'd let go of the rope in the split second between hitting the wave and being trapped upside down in the swirling water. With nothing but an old, well-worn life jacket keeping me partially afloat, down the rapid I went.

Despite the trauma of the situation, I remember it all so well.

I remember spinning constantly, not knowing which direction I was facing. I remember being pulled underwater into the darkness, and occasionally floating close enough to the surface to see the steep, high cliffs and the sun flickering above me. Once or twice I was able to take a big gulp of air. It's amazing how far those breaths can go when you don't know if you're going to get another.

I remember water filling my ears, and the incredible noise it made as it rushed by, hitting rocks and swirling and crashing back on itself. My senses were all over the place and I couldn't process anything as I was dragged violently downstream. I remember, at one point, putting my arms above my head, my hands occasionally breaking the surface, as I desperately hoped to grab onto something to hold – anything to end the nightmare. The torture seemed to go on forever, even though it only lasted a minute or two. When your life feels in danger, and your fate is out of your hands, that's more than long enough, I can tell you.

Then I remember a sudden calm coming over me, one that's almost impossible to describe. I stopped struggling, resigned to my fate. I no longer felt scared. Incredibly, I just decided to go with it, not to fight any more and just to let whatever was going to happen, happen. A single thought came to me, out of the blue, clear as day, that I'll also never forget: I'm not meant to die this way. If that's the case, I thought, then why bother struggling? Why worry about this being the end? How on earth I'm meant to die I'll never know but, with that one fleeting thought, my mood changed completely.

Eventually the noise stopped and I ended up on my back, gently floating in calmer waters. I took my first proper breath as the bright sun slowly began to warm my face. As my eyes and ears cleared I could see everyone else in the distance, by this time back on the raft, looking my way with anguished looks on their faces.

'We thought we'd lost you, Ken!' one of them said as they gently hauled me back onto the raft.
'Me, too', I spluttered, water still dribbling out of my mouth, ears and nose.
'Man, your hands rising up out of the water like that. We all thought that was it.'

There were still a few rapids left before the scheduled lunch break and I stuck with it, partly because I thought nothing could be worse than what I'd just been through, and partly because you can't just get out of the raft any time you like and leave. When lunchtime did come I was asked whether I wanted to call it a day – I could quite easily have walked back from there – but there were only a few rapids left and, crazily perhaps, I felt a need to get to the end. So I stayed and, despite a couple of final scares, we all made it safely back to the hotel complex that evening with plenty to talk about. I remember not eating or drinking much for about a day, though, and having an awful stomach ache, almost certainly caused by the volume of water I'd swallowed during those two eventful minutes.

Our last couple of days in Livingstone passed without further incident. We walked across Victoria Falls Bridge, a favourite bungee jumping spot for other adrenaline junkies, and into Zimbabwe to get our passports stamped, and then walked back. It all seems pretty pointless now, but everyone seemed to do it back then. Over a decade later I'd visit Zimbabwe for real as my mobile technology work took off, and camp on the banks of the Zambezi, the river that almost cost me my life. We spent time browsing and haggling with the stall holders selling paintings, carvings and other souvenirs of the Falls, and bought as much as we could carry, bartering away everything from our shoes and socks to our sunglasses and hats. We had one last evening drinking beer in the shadow of the Falls and, two days later, we were back home, back to our regular lives in a totally different, privileged world.

I'll never forget the culture shock I experienced when I returned to Jersey, and I felt terribly uncomfortable with how easy my life was. Electricity came on with the flick of a switch. Crystal clear water flowed with the effortless turn of a tap. The shops were piled high with brightly coloured fresh food. The roads were tarmacked and smooth. Street lights came on at night. Traffic lights changed with monotonous regularity. Everything, in fact, just worked. How different it all felt from the place I'd just been.

But one of my biggest disappointments was the ease with which most of my team members seemed to slot straight back into their regular lives, without too much thought or struggle. Was I just being oversensitive, as always? Whatever it was, I never forgot the people I'd met in Zambia, and the poverty and struggles they faced each and every day. I was determined that Chilubula was going to be the start of something for me, not the end.

Despite wanting to do more after we got back from Zambia, I returned to work with Vic, and continued my coding at Jersey Zoo. I did keep reading, kept watching, kept paying attention, and kept trying. I applied for another Jersey Overseas Aid trip a year later, but was turned down. 'We want as many different people as possible to get the experience, Ken' was their reply. I then applied for a two-year placement with Voluntary Service Overseas, and attended an interview and orientation session at their head offices at Putney in south-west London. Not only did they turn me down but they also told me they thought international development wasn't a good fit for me. What little confidence and enthusiasm I had were being well and truly put to the test now. After a glimmer of hope in Zambia, it was back to that all-too-familiar feeling of rejection and disappointment.

Zambia never did feel too far away, though. As promised, Justice and I kept in touch and I regularly sent him cash and artist's materials to help with his painting. We exchanged letters, which I still have stapled to new sketches he'd also send me. His lion painting hung proudly on my wall in Jersey, as it has in all my homes since.

After about a year the letters stopped, though, and I got worried. I checked in with Kevin and Vanessa, who faxed their friends in Chilubula to find out what was going on. A couple of days later the phone rang, and I picked it up.

'Hi, Ken, it's Kevin.'

'Oh, hi Kevin. What's up?'

'We've got news back from Chilubula. I'm very sorry, but Justice has passed away.'

I paused for a moment. I was half-expecting the news to be bad, but it was still a shock.

'Do you know how he died, or when?'

'I'm afraid not, Ken. But we think it was a couple of months ago.'

'What's happening with Chiti and his brothers?' I asked.

'No one knows yet. It's all a bit up in the air as they figure out what to do.'

My first instinct was to worry about his sons and what would happen to them. They'd already lost their mother, and Justice had worked so incredibly hard bringing them up all on his own. I hated the thought that all his efforts would go to waste – there was already talk of the children being split up and sent to live in various church homes –

and without money they'd all be forced to drop out of school. They had no one left now and I felt a huge burden of responsibility, thanks to my friendship with Justice, to do something to help them.

I immediately started thinking of ways to raise money. I knew how much I'd need to keep them all in school, and just had to find something extra to keep them together in the same home. There was talk of building a house, but we'd need even more money for that. The first thing I did was purchase wood, and started making replica copies of Justice's lion painting. I only sold a few of those, but it all helped. I wrote begging letters to the local paper, and reached out to various local charities, including the Rotary Club of Jersey, for donations. An old friend whom I'd gone to primary school with gave me permission to hold a barbeque on his family's farm, and I got free drinks from a local brewery for the event. Nigel agreed to bring his band along again, and a few dozen people bought tickets. Collection tins spread around shops on the island topped it all off and, in the end, I had somehow raised a couple of thousand pounds that I sent off to Zambia. Chiti and his brothers all did finish school, and they did all stay together. Justice would have been proud.

Today they're all grown up, and happy and healthy. Chiti paints and draws, exhibiting a natural talent he inherited from his father. His brothers took more of an interest in mechanics and, today, they run their own garage while Chiti runs his own visual arts and print shop. I still send money to Zambia, but now it goes to Chiti to help with materials and equipment, and we have become close friends.

It took just two trips to Africa, Zambia in 1993 and then Uganda in 1995, to seal my fate. Strangely, I remember little about that second trip with Ray, so there's little I can share here. Confused, concerned and feeling guilty about much of what I was seeing and reading, and without any obvious way of helping fix any of it, I decided the best thing to do would be to go to university and study global development. There wasn't much keeping me at SuperCard, where I now worked, or on the island, come to that. And even though my job was going well, I knew it wasn't a long-term option for me. Most days I was simply going through the motions.

In part, my desire to go to university was a means to an end. Most humanitarian jobs I'd seen advertised required a degree of some kind, so I thought it best I try and get one, despite my generally abysmal academic record. There were no universities on the island, so that could only mean one thing.

I'd have to sell up and leave.

The Pier

Its fall into the sea was almost complete when I last stood there in the summer of 2010. I had been invited to give a keynote speech at an event being held in a hotel on Brighton seafront and, as was always the case when I was in town, I took the opportunity to take a leisurely walk along the sloping, pebbled beach towards Hove. Two fires in 2003 – both deliberately lit by some accounts – were the nail in the coffin for any fleeting attempts at restoration. The very first time I was there, about 14 years earlier, there was still just enough of the steel skeleton remaining above sea level to present a vivid picture of what would once have been a showpiece Victorian structure. It was grander, by all accounts, than the Palace Pier a little further along the seafront, which still stood proudly, big wheel spinning, children screaming, slot machines whirring, lights flashing. On what remained of the West Pier, adults, children and brass bands had long been replaced by roosting pigeons, and starlings that burst intermittently through large holes in the dilapidated roof, swirling around in the sky above the wreckage, returning a few moments later after a mesmerising display of synchronised flight.

Despite all the destruction, the scene was strangely beautiful.

West Pier was constructed during the Victorian tourism boom of the 1860s, driven in part by an obsession with the health benefits of sea air, and was one of an incredible 22 piers built around the country that decade. Improvements in rail travel – things like better comfort for passengers and faster journey times – had opened up the south coast of England to mass travel for the first time, and Brighton was one of a number of seaside locations to reap the benefits. The National Archives believe that over £3 billion was spent building out the railway network in the second half of the 19th century. In 1870 alone, a staggering 423 million journeys were taken on the trains. By the end of Queen Victoria's reign, over a billion journeys were being taken for business and, of course, pleasure. Trains were a big deal.

Influenced in large part by this new-found mobility, construction of the West Pier was completed within three years at a cost of £27,000, despite resistance from a number of wealthier residents who feared it would ruin their views of the sea. It was opened, to much fanfare, on the 6th of October, 1866, by Alderman Henry Martin, the town's Mayor. Research into my ancestry was to later reveal that Henry Martin was, in fact, my great-great-grandfather, and that Brighton was insanely rich in family history. The city was also to play a major part in my own journey, but more of that later.

Once open, West Pier proved enormously popular, boasting millions of visitors either side of the First World War. Just a few short years after my great-great-grandfather had cut the ribbon, expansion plans were already afoot thanks to growing visitor numbers, and the pier was extended to allow for the addition of a new pavilion. Opened in 1893, it boasted an impressive capacity of 1,400 people. A little over two decades later, in April 1916, a new concert hall was added as competition for visitors heated up between West Pier and the other local attraction, Palace Pier, which had opened in 1899 a few hundred metres along the shore.

I remember the very first time I stood next to what remained of West Pier. It was one of those dull, overcast, blustery kind of days that Brighton seafront seemed made for, and the conditions perfectly matched my mood. It was early October 1996 and I was thirty years old. Sussex University campus, a short rail journey away in Falmer and my home for the next year, was already awash with red, orange and brown as most of the leaves parted ways with their summer hosts. I'd made the first big decision of my life over a year earlier, and the upheaval I now found myself in the middle of made the last four months of 1996 a particularly challenging time. Every year since, the smell and feel of early autumn reminds me of how lost I felt. I feel it to my bones. Anyone who suffers from Seasonal Affective Disorder – more commonly known as SAD – will know it well. The onset of autumn is my unwelcome annual reminder of what I put myself through back then.

I'd left my last job at SuperCard and sold everything I owned two months earlier to go to university, and now I wasn't even sure if I wanted to be there. They wouldn't have realised at the time, but the many doubters at home who told me I'd be back within a few weeks were the ones who drove me on to stick with it, despite my own considerable doubts, worries and reservations. Stubborn is a word that springs to mind and, on this occasion at least, I was grateful for it. No way was I going to quit this early.

I wasn't the only one struggling with the adjustment. As a mature student I was drawn into many of the social groups dedicated to helping older learners settle in. Like me, many had given up full-time, relatively well-paid jobs and homes of their own. Unlike me, some decided early on that they couldn't cope, and quit. For a while I found myself in the unenviable position of offering emotional support and encouragement to other mature students when I could have done with a dose myself. In some cases I succeeded, and in others I failed. I still wonder what happened to those who decided to call time early on their university careers.

I was thrilled when I received a letter from Sussex University earlier that summer informing me that I'd secured accommodation on campus. Foolishly, I thought that living there would make my transition easier, and remove travel from my long list of things to worry about. How wrong I was. I found myself living in some of the older university accommodation in a small, nondescript room on the far side of campus, sharing Block 68 in East Slope with eleven other people, most of them students in their teens who had never lived away from home before. For them it was a brand new, exciting adventure, a new world of freedom and independence. It was a step up, not a step down as it was for me.

For most of my housemates it was also a seamless transition from college to university. I, on the other hand, hadn't been anywhere near a classroom for 13 years and had never researched or written a proper essay before in my life. For a long time I felt out of my depth, and unsure of where I was, and I doubted whether I had it in me to succeed at this level. It didn't help that I found campus life bland, clinical and suffocating, a microcosm of the kind of small island life I'd only just escaped. I spent a lot of my time on trains in those early months, heading mainly to London and Brighton, particularly at weekends when many of my housemates had the luxury of popping back home to see their families or friends. I could never do that. Campus was a hard enough place for me when it was busy, but it was something else altogether when it emptied. I really couldn't win.

Within six months, my initial excitement had completely vanished and I decided I'd had enough of campus living. I called it quits and moved into a one-bedroom flat in the centre of Brighton, desperate to try and recreate at least a little bit of the stability I craved. It was small but comfortable, a little too traditionally furnished for my liking, but it gave me the space and independence I badly needed. I now had to 'go' to university, giving me the kind of separation I was more familiar with in my Jersey days when I had to leave home to get to work. It also gave me time and space to be alone, something I've always needed but foolishly neglected to factor in with my initial desire for campus life. The flat was just off North Street, around the corner from all the main bus stops, close to the main shopping centre and a short walk from the train station. Just as importantly, it was only a hundred metres up the road from the seafront where there were plenty of takeaways to choose from when I felt lazy (or was in need of a treat), and more than enough cafes to work from. It was a move that not only rescued me mentally, but also breathed new life into me academically.

Thankfully, the first year of my Sussex degree was foundational, so not only did it not count towards my final mark but I only needed a modest pass to progress to year two. While none of my essays or exam results would go towards my final assessment, my marks would obviously be an indication of how well I was adapting after a challenging start. This first year also gave me time to learn how to write and research properly, and how to structure essays and papers academically. I really didn't have a clue when I first arrived, and it showed in my early pieces on colonialism and development. Not only was the depth of the subject matter new to me, but I had to learn how to explain myself in ways that I had never done before, structuring my arguments logically and citing key facts and sources. Still, the lecturers were patient and could see I was genuinely interested in the subject matter, and that I was trying my best. It probably helped that I was one of the strongest in class and relished the kind of in-person debate I was starved of back home. My written work gradually improved, putting me on course for a respectable upper-second class degree, assuming I could maintain the momentum.

The summer of 1997 came quickly, signalling the end of my first year and the arrival of increasing numbers of tourists to Brighton and its seafront. After the most challenging of starts I'd somehow navigated my way through a dozen pieces of written work, two exams and, more importantly, a huge amount of self-doubt. The second year lay ahead when the serious work would begin. But for now, a much-deserved summer break beckoned. Living in the centre of town, close to the beach, cafes and bars, put me in a great place to enjoy it. For the first time since leaving Jersey, I was in a good place mentally, too.

That summer would also prove to be significant for other reasons. While working as a part-time activities organiser for a foreign language school, I met a girl from Finland. Exactly ten years later she would become my wife, and Elina and I would marry a short walk away from my town centre flat, in the Royal Pavilion building – an incredible piece of architecture which wouldn't look out of place anywhere in India – and a place in which Henry Martin would have regularly hung out. Brighton, as my relatives found, seemed to be a place that just kept on giving.

University soon started up again, and things once again took a turn for the worse. I struggled to get my head around any of the early social anthropology or development studies modules, and hit Christmas with new bouts of self-doubt.

I wasn't just imagining the struggle. I had lower marks to prove it. But it wasn't all bad. I'd called time on my flat and moved into a shared house in Hove with a few friends, bought myself a small scooter to help me get around, and found a part-time job in the computer department at Varndean, a local sixth form college, where I also taught evening classes. Life outside university was good and I was back earning money, enough at least to protect some of my ever-dwindling life savings.

I'd been on something of a rollercoaster ride since arriving in Brighton, but as the troublesome first term of my second year came to an end, everything once again began to look up. Starting in the new year I'd be able to choose my own social anthropology and development studies topics, so I'd no longer be forced to study things I didn't like or understand. With my new-found freedom I started enjoying my studies again, even though I still needed convincing that I could perform consistently enough to succeed at university level. My first piece of assessed work – a study of cross-cultural interpretations of violence – gave me the boost I badly needed. I spent more time on that 1,000-word concept note than I had on any piece of writing in my life, pushing it through multiple edits and seeking advice from anyone who showed an interest. It was a subject that genuinely interested me though, and, more importantly, made sense to me. I had decided at the start of that second term that I wouldn't take on any new modules that I couldn't, in some way, relate to.

After a huge amount of effort, I finally handed in my first piece of assessed work. Then I waited. And waited.

Three weeks slowly passed.

I was nervous as I headed to the school office that lunchtime. I'd been there two days earlier, only to be told that the tutor was yet to hand in the marked papers. I pressed the white button on the counter and heard a faint buzzing sound near the back of the room, followed by the sliding of chair legs across the floor. A few seconds later, the clerk appeared.

'Yes, what can I do for you?' he asked.

'I was wondering if the second-year social anthropology extended essays had been marked and handed back in yet?'

'Oh, yes, you came a couple of days ago, didn't you? Let me check.' He disappeared and came back a couple of minutes later with a scruffy brown folder full of papers.

'Name?'

'Oh, it's Ken Banks', I replied, nervously. He flicked through the folder and pulled out my work.

Second time lucky. My paper was slid to me across the counter. I didn't look straight away – that would have seemed desperate – so I said thank you and walked casually around the corner, stopping halfway down an empty corridor. Holding my breath, I turned the paper over and unfolded it to look at the top sheet, the yellow one stapled by the tutors with their comments and overall mark.

To say my heart raced would be an understatement. I had to look twice. A 'first'! I'd got a 'first', and quite a comfortable one at that. (A 'first' is the highest category of undergraduate degree you can get at university in the UK. Any mark over 70% falls into this top tier.) I'd not managed to get anywhere near 76% in any of my previous work, and the achievement hit me like a bolt of lightning. I was dumbfounded and hugely proud. Who would have thought that a hastily written number, and a few words, could have such an impact on a person. I had to remind myself that it was only one piece of work, and one worth just 10% of my degree, but for the first time in my life I finally believed I could achieve something academically. It was a belief that kept me going for the rest of my days at Sussex.

I put the papers in my bag and scuttled off excitedly to phone my mum.

Things continued to go well and I continued to enjoy my studies. I looked forward to the seminars and lectures and no longer dreaded picking up my marked essays. By the time the summer term ended I was on course for a 'first' in both social anthropology and development studies. It was some turnaround. I carried the official piece of paper listing my marks everywhere with me that summer, and eventually started using it as a bookmark. Few days went by without me taking a glance. For the academic underachiever that I was, it was the best pick-me-up imaginable. That small piece of paper had 'belief' and 'you can do it' and 'everyone at home is wrong' written all over it.

My break that second summer was a little unusual. A few months earlier I'd come across an organisation looking for volunteers to help with a three-month biodiversity survey in Uganda. It was one of those adventure-style trips, one where you're expected to pay a not-insignificant amount of money to carry out scientific research, research that usually turns out not to be scientific enough to be that useful to anyone. I'd already met my half-dozen team members during an orientation weekend in the spring, and had managed to cobble together the fee through a combination of part-time earnings at Varndean, previous savings and a chunk of paid work helping with the computer systems at Jersey Zoo over the Easter break. It was a three-month trip, my longest to date, that would take up the entire summer break. Fortunately I'd hit my stride and didn't need to pack in any extra study before my final year, so I could drop my guard and relax for a while. And yes, I did take my bookmark with me.

Karuma Wildlife Reserve, our destination for those three months, is a 280 square mile area straddling the south-eastern border of Murchison Falls National Park. Named after the intrepid Victorian explorer and geologist (and former President of the Royal Geographical Society) Roderick Murchison, it was briefly renamed Kabalega Falls in the late 1970s during the reign of Idi Amin. From a biodiversity perspective the main action was in Murchison Falls, and Karuma acted more like a buffer zone, keeping any initial encroachment as far away as possible from the wildlife and habitat that needed most protection. A long, winding, tarmac road ran along much of the edge of the Reserve and it was there, 15 minutes' drive down a dirt track opposite Nyamahasa Primary School, that you'd find our main camp. We spent time there resting and catching up with washing and general maintenance in between forays deep into the Reserve where we'd set up satellite camps for a few days at a time. Satellite camps allowed us to carry out biodiversity surveys at strategic points around the Reserve, with the aim of helping the authorities decide how best to manage the area.

Life was often remote and, because of that, incredibly simple. Fortunately we all got on, which is critical when you're all stuck together in the middle of nowhere. When we weren't out doing bird counts, checking traps, measuring tree density, or cutting or walking transects – paths through bush or dense vegetation, often cut with large machetes – we'd be cooking, playing Uno, chatting or reading. *Into The Wild* by Jon Krakauer was a particular favourite of mine, and a set of *Where's Wally?* books I'd brought with me became everyone else's favourite way of passing the time. Sometimes we'd just listen to music when there were batteries available. I had a cassette of George Michael's album *Older* and that became our signature music for the trip, not to everyone's pleasure, I have to say. One song in particular, 'The Strangest Thing', still reminds me of nights we spent sitting out in the darkness, lying on our backs in the long grass, taking in the sounds of insects as we watched the flashes of faraway storms and the flickering of distant stars. There was something magical about being so far away from civilisation, so close to nature, and about living in the most simple of ways. I miss that, if I'm honest.

Our main camp had communal tables and chairs made out of locally cut wood, a shower cubicle (but no actual shower) and a washing and cooking area. It was precisely the sort of camp you'd expect to find in an uninhabited wildlife reserve, made speedily and with the scarcest of tools and materials. But it was home, and a place we could recharge and rest. Our satellite camps, on the other hand, didn't have the luxury of any fixed furniture and, on arrival, we often had to pick a spot, clear the ground and make the best sleeping, cooking and washing arrangements we could. After a long day of travelling, followed by a backbreaking hike carrying our equipment, having to make a camp was the last thing any of us usually wanted to do.

The cooking and washing areas were generally the easiest to organise. Just a few dozen slashes with our machetes to clear the ground often did the trick. The sleeping area was something else, though. For a start, we'd need a couple of sturdy trees, the right distance apart, to tie the supporting rope between, and finding those could be challenging, particularly in the savannah. We then needed flat enough ground, which we had to clear before laying down our blue UN-branded tarpaulin. Another tarp was thrown over the rope and pegged down on the ground, then stretched across at an angle so that it covered our sleeping area. It was the simplest of designs that resembled a large wedge of blue cheese. Apart from the hard ground it was comfortable enough, and it kept us and our clothes dry even in the worst rainstorms, and we had plenty of those.

We slept side by side, from one open end of the tent to the other. We could all just about fit in with our bags pushed down near our feet. Matt and Roger slept at the ends, something I was grateful for. We joked about who would be picked off first if any hungry carnivores came our way in the night, and the ends were obviously the most exposed. There were no real scares to speak of during all our time in these satellite camps but one morning, at main camp, we did find a puff adder exploring the underside of our makeshift tent. Nobody had a lie-in that morning, I can tell you.

Snakes weren't the only thing we had to contend with. Ants gave us by far the most trouble. There were aggressive ants in all shapes and colours and sizes with two things in common – an incredible bite and a terrible attitude. Early one evening, as we headed back to one satellite camp in the fading light, we unknowingly brushed past undergrowth teeming with tiny ants, hundreds of which ended up on our clothing. It took a few minutes before we realised but, once we did, in no time at all we were all jumping around in fits of frenzy, battling in vain to flick and pull and brush them off our bodies. These ants may have been small, but they packed a mighty punch. On one other occasion, this time at base camp, we had to dig into our paraffin supplies to set fire to the ground to clear a huge swarm of large brown ants that had the nerve to embark on a dusk raid. We all stood on the tables and chairs that time, flames shooting up all around us, witnessing the destruction in disbelief.

We also had an eventful couple of days in another camp fighting off bees, this time after the park rangers travelling with us discovered, and then raided, their nest. Honey was too good to pass up, so they decided to climb the tree, scoop out loads of wet, dripping, shiny honeycomb with their bare hands, bringing it back to base with them, eating most of it along the way. Swarms of angry bees followed them back in hot pursuit. Yes, the fresh honey was delicious, but we nearly paid a price. Fortunately – amazingly – none of us got stung.

As we approached the half-way mark in our trip, we decided we deserved a little time off. After six weeks of the most basic living, with constant ant attacks, no running water, no electricity, no proper beds, and without any real rest or relaxation to speak of, a few of us were up for a short holiday in Jinja, a city about 50 miles east of the capital, Kampala. I thought we'd sit around in the sun, swim, read, drink beer, eat food that wasn't beans and rice, sightsee and generally be tourists for a while. I, for one, couldn't wait.

'So, who's up for Jinja, then?' I asked, hopefully. There were one or two nods, but not the unanimous signs of enthusiasm I was hoping for. Instead, Roger leaned forward and glanced over.

'I don't know about any of you, but I wouldn't mind doing something a little more adventurous. Maybe climb a mountain or something.' I kept quiet, eyes fixed straight ahead, expecting others to come to my rescue and say what a stupid idea it was. But nobody did. I couldn't believe anyone would want to go and rough it even more, this time up a mountain, after weeks of roughing it, when a chilled week was the alternative.

'What did you have in mind, Roj?' asked Matt.

'Well, if we head to Kisumu, we could climb Mount Elgon from there. It's about a day's drive to Kisumu from here.' I remained quiet. I knew a lost argument when I heard one.

So there you have it. We were going to spend our only week off climbing a mountain.

Mount Elgon sits close to the Ugandan border with Kenya, and is estimated to be around 24 million years old, making it the oldest extinct volcano in East Africa. It has five peaks, the highest coming in at 4,321 metres above sea level. That peak, Wagagai, is entirely on the Ugandan side (some of the peaks cross over into Kenya) and Wagagai was our destination. The early stages of the hike were pleasant enough, with lush green vegetation and freshwater streams the order of the day, and little by way of incline. It was peaceful and strangely relaxing. But that soon changed as we started rising in elevation, gentle slopes turning into steeper hills. And then, just a few hours in, we came across our first major obstacle. The 'Wall of Death'.

If you head up Mount Elgon today you'll find a fairly presentable set of wooden steps which gently take you up the 20-odd metre rock face that we found standing in our way. But not in 1998. It may not have been 20 metres – I can't exactly remember, to be honest – but it was quite an obstacle and we climbed it vertically, grabbing onto pieces of loosely-nailed wood cobbled together to make a sort of makeshift ladder. The 'Wall of Death' did sound like an appropriate name at the time, and if any of us had fallen we'd have been in serious trouble. Suffice it to say, we all made it up, one at a time, slowly and gingerly, without any of us looking down. As it turned out, going up was much easier than the descent we would face a couple of days later.

After a chilly night's sleep in one of the overnight camps on the mountain, preceded by an instantly forgettable meal of plain white rice and freshly chopped bamboo (there's plenty growing on the higher levels of the mountain), we set off early the next morning for the summit. It wasn't a huge distance – Mount Elgon isn't really that high in the grand scheme of things – but as we edged our way up we all became noticeably slower and increasingly out of breath. Breaks became more frequent, sometimes after just a few steps. I'd never been at this altitude before and was amazed how hard it was to walk as we crept ever closer to the 4,321-metre mark. We did eventually make it, all of us, and we took photos to prove it. This might not have been the relaxing break I was hoping for but, at that moment, I did feel an immense sense of achievement and, although I never admitted it to the others, I was glad we'd decided to do it. It was to be the only mountain I'd ever climb. A few years later I was offered the chance to work on Mount Everest with National Geographic, an opportunity I had to turn down as we awaited the birth of our first child, Henry.

Our descent was fairly straightforward, apart from the 'Wall of Death' that is, which was a little harder navigating in reverse. But, once again, we slowly inched our way down, one by one, showing it the respect it deserved. During the hike from there, back towards the village, we broke up into smaller groups. Some of us were fitter than others, and some were just desperate for a cold Coke. Others just wanted to enjoy it and take their time. Andrew, a hippy-looking, bearded Rastafarian team mate from Scotland, ended up walking with me and we were met, just outside the village, by a welcoming party of young children and an assortment of dogs. The children cheered, running alongside us, as we took on the last couple of hundred metres of the dry, dirt road, before reaching our final destination.

Our adventure didn't take up the whole week, so we did have some time left to relax, play pool, eat good food and have the odd beer. I remember having severe yoghurt withdrawal symptoms and would often substitute a cold beer for a one-litre bag of locally-made strawberry yoghurt. Because I was used to drinking a lot of milk back home – and, yes, eating my fair share of yoghurt – this was one of the few things I'd really missed.

The rest of our time in Uganda took the same pattern and routine as the first half, although we did manage one final, short break, this time to Murchison Falls where we sat in a boat, in awe, just a couple of hundred metres from the crashing, cascading, bubbling mass of water. The Falls release over 300 cubic metres of the stuff per second, all through a gorge less than 10 metres wide. Victoria Falls it may not have been, but it was quite something to see, and quite something to be allowed so close to.

As we approached September it was time for us to leave. Andrew and Heather, who worked for the organisation, were staying on to meet the next group who would be coming in a couple of weeks later. We had the obligatory leaving party that involved way too many warm beers long into the night, around a roaring campfire, to the sounds of Fred Sebatta, a Ugandan artist I'd begun to like after hearing his music blaring from a local bar (I still have the cassette tapes, even if I don't have anything to play them on). It was obvious that my fellow travellers were grateful not to be subjected to George Michael one final time, particularly on their last night. All-in-all, it was a sad farewell. We had become a tight-knit group over those three months, an unlikely band of lawyers, accountants, sailors, computer engineers and environmentalists. We did stay in touch and for an incredible eight years we had annual summer reunions and get-togethers. Marriages, mortgages, careers and children finally put paid to those.

As soon as we arrived back in the UK we said our goodbyes and fanned our way back out across the country, different destinations and different destinies awaiting us. I was headed back to Brighton and a house on Aberdeen Road that I'd be sharing with a mixture of friends from my first and second years at university. I'd not seen the property, but they had secured me a room while I was away, and it was in a great spot, slap-bang between Brighton town centre and university. On the home front it was to be a great year, by far the best I had in any house.

Things now began to get serious with my studies. One of the first modules of that final autumn term, Linguistic Anthropology, looked at the role of language in the social lives of communities, and there was a certain logic that drew me to study it. Jersey used to have its own unique language called Jerrias (or Jersey French), and it was dying out as many of its speakers aged and passed away. There was an increasing effort to try and preserve it in some way. Similar movements were springing up in other places, such as the Isle of Man, and within other indigenous communities around the world. Following my earlier decision to focus only on things that interested me and that I could relate to, studying something closely connected to my home island made a lot of sense. As it turned out, I enjoyed researching Jerrias (and the wider topic of language loss) so much that I had all 4,000 words fully researched, written, edited and ready to go by the end of the Christmas break. None of my friends could believe I'd finished one of my main essays six months early, but it turned out to be a great move, one that allowed me to focus the rest of my time and attention on other, more challenging pieces of work.

As other modules opened up, so did opportunities to write and research other subjects that interested me. Being able to choose was a blessing, and led to my final development studies' essay focusing on the concept of 'appropriate technology' in the developing world. I didn't realise it at the time, but E F Schumacher and the wider appropriate technology movement would have a huge impact on my future work. Ironically, it would also lead to World Watch Magazine describing me years later as 'possibly the world's leading voice on the use of mobile phones as an appropriate technology'. To this day, I remain intensely proud that I used everything I learnt at university in my later career, and for that reason alone I'm grateful that I managed to stick it out. Oh, and that I proved all the doubters in Jersey wrong, of course.

The last few months at Sussex went quickly. Timetables emptied as lectures and seminars ended, and everyone hunkered down to write their dissertations or revise for their exams. For the next few weeks the pubs witnessed none of the usual student hustle and bustle. Bedroom lights instead burnt long into the night, and assortments of takeaway boxes piled up in bins outside front doors, overflowing into the street. It was all so very different from the previous two-and-a-half years. Suddenly, things had got serious, and hundreds of students, me among them, found themselves at the business end of their university lives.

Things went particularly smoothly for me. After everything I'd been through, I felt I deserved it. Having completed my Jersey French essay several months earlier, I was far more relaxed than I would have been. Fate decided to play it out that way, and I was grateful for it. My mind, and my time, were free to dive into my growing passion for anthropology and conservation, and my main dissertation focussed on the role anthropologists played in the creation of protected areas around the world. Sadly, despite their involvement (and, in some cases, because of it) many indigenous communities found themselves moved away from land deemed critical for wider wildlife and biodiversity conservation efforts. The impact of those decisions, many made a century or more ago, are still felt around the world today.

I made contact with some inspiring anthropologists and ethnobotanists during the course of my research. One of these, Michael Fay, became famous for hacking his way through the Congo Basin on a 2,000-mile, 465-day 'mega transect' in 1999. As fate would have it, I'd meet Michael, along with many other conservation and explorer heroes, just over a decade later when the National Geographic Society picked up on my own work.

With essays complete and handed in, I now turned my attention towards my two examinations. The last time I'd taken exams that really mattered I was in my mid-teens, and I failed most of them, due in large part to a lack of interest in the subject matter if I'm really honest. It may have been 16 years since I last felt the pressure of an exam room, but this time there was one notable difference. I was fully engaged, excited and inspired by what I was learning and, as I discovered, when my brain got switched on to a topic it became sponge-like, soaking up everything in its path.

I did my research and tried to second-guess at least three of the topics the exams might cover. Focussing my attention on those, I wrote around four pages of bulleted, summary notes for each, which would be relatively quick and easy to read over and over again until I could recite it all in my sleep. I was already familiar with the topics I'd written for my dissertation and extended essay, so I was confident I had it all covered. The first thing I'd do when I got into the exam room, I thought to myself, would be to pick the three questions I'd answer and quickly write down as much as I could remember from those summary lists.

When exam time came, my plan worked a treat, and I came away from both feeling that I'd done the best I could. And I ended up with the marks to prove it.

'How did it go then, Kendo?' asked Phil, one of my earliest friends at university, as I returned home. For some reason Phil, and most of my other friends, decided to call me Kendo. We even named our five-a-side football team, of which I was the captain, Kendo's Killers.

'Okay thanks, Philip. I was able to answer three questions so that's always a good sign. Complete and utter disaster averted, I'd say. Now the waiting begins!'

'You'll breeze through, Kendo. Got time for a celebratory pint? I need a break from my geography revision.'

'Happy to help, Phil', I laughed. 'You paying?'

All of my friends ended up doing well when the results came out a few weeks later. There were a few surprises, in particular a couple of shockingly high marks from people who thought they'd done poorly. I remember results day well, even to the point of the music I was playing as I took the long walk up to campus to get mine. 'Say Hello, Wave Goodbye' by *Soft Cell* became my pre-results mood music, and it still reminds me, to this day, of the anticipation I felt as I took that 45-minute walk in the summer sunshine. I must have played it nearly a dozen times before I got to the famous concrete 'University of Sussex' welcome sign that particular day.

Close to three years earlier I'd arrived at Sussex with very little by way of expectations, and I very nearly fell before reaching even the first hurdle. Perhaps success for me was simply being there at the end. Of course, no employer would see it that way. Perhaps just passing with any kind of half-respectable mark would be a success of sorts, but would all that financial and emotional cost, upheaval, pain and frustration be worth it for just any old mark? Or, given I'd exceeded expectations and achieved a 'first' in my second year, would anything less than a 'first' be a failure? For an outcome now in the lap of the Gods I was doing a great job of stressing, sweating and obsessing over all the possibilities during that results-day walk.

I got to university early and decided to go to the gym, one last time, before heading to the corridor of doom where the results would be pinned to the wall. I'd managed to go regularly during my entire stay at Sussex, at times feeling like the physical punishment I put myself through was the only thing keeping me balanced and sane. I was in good shape when I left Jersey, thanks to a new-found obsession with swimming, and I was keen to try and keep it that way. I did my usual workout that morning in the gym, which turned out to be a little busier than usual. Clearly I wasn't the only one wanting to take their pre-results nerves out on a rowing machine or bench press.

I left the banging and clattering of metal, and the heaving and puffing of bodies, and took the short walk across campus to Arts C, the building which now held the keys to the futures of hundreds of ambitious, proud, eager students. Maybe all this mattered more to me considering everything I'd given up to be there, and my previous lack of academic achievement. As I took each step, I couldn't help but think back to that warm, Saturday morning in summer 1982 when my secondary school exam results arrived in the post. I was in bed, probably writing a poem or something, when Mum sheepishly walked into the room clutching an envelope.

'Do you want to open it, or should I?' she asked.

'Huh? What is it, Mum?'

'It looks like your 'O' Level results.'

'Oh', I replied. The day of reckoning was finally here. 'Do we have to?'

I let her open it. I could tell she wanted to. I'd struggled at school, and my final exams had coincided with the FIFA World Cup, held in Spain that summer. I'd watched too many games in a desperate bid to avoid revision. Letting her open them would delay my agony just a fraction longer, I thought. As she scanned the sheet of paper, I watched her face. There was no big smile, no whoops of joy. I ended up scraping through in just two subjects, English and maths, and failed six others. To say it wasn't great was an understatement. Back then I didn't feel it was going to make much difference in my life so, while I was disappointed, I wasn't overly upset or surprised. This time, however, as I approached the double swing doors of Arts C and began taking those last few steps up to the second floor, things mattered a whole lot more.

I was so early there still weren't that many people around. That was good, I thought. Fewer people to witness my disappointment. A few students had already seen their results and they were heading back in the opposite direction, some with smiles, some without, and some with expressions that were impossible to read. Within a couple of minutes I would be among them.

Nothing – quite literally, nothing – had been straightforward for me at university, not to mention those frustrating few years beforehand which finally led me to take that leap of faith and go. If you thought the simple act of picking up my university grade was going to be any different, you'd be wrong.

After my tremendous set of second-year results, my expectations had hit levels that could only be described as unrealistic. I felt I'd worked twice as hard in my final year, so it didn't feel crazy to believe that a first-class degree was a strong possibility. Not only would a 'first' make up for all my years of academic underachievement, but it would also have been the perfect two-fingered salute to everyone who had said I'd fail again. And, more importantly, it would make my mum proud. Despite my age, making her proud was still so very important to me.

I got my results. And I missed out on that 'first'. By about three-quarters of a percent, apparently. I don't think it's possible to get any closer and not creep over the line. Taking everything into account, a very strong upper-second class degree was a huge achievement, but it didn't feel that way at the time, or for a very, very long time. As we approached graduation day, random conversations with faculty revealed just how close I had been. They'd apparently discussed reviewing my grade. If just one of my essays or exams had been a tiny one percent higher, I'd have made it. So, yes, they had discussed it, and fretted over it, and debated it. And then had decided not to do anything about it. I was gutted. Totally gutted.

There was another, more important reason for my deep disappointment and frustration. I'd started to think about sticking around to do a master's degree, and taking a more specialist line in anthropology or development studies, or perhaps a combination of the two. But competition for funding was fierce, and I knew I'd need money. Students with first-class degrees, which back then accounted for as little as 5% of them, stood out straight away and had much greater chances of financial support. The bucket I now found myself in represented around 45% of students, so I was a small, averagely intelligent fish in a great big sea. It just wasn't going to happen, so I didn't even bother trying. My master's plans were shelved and didn't re-emerge until a couple of years later when I found myself in Finland.

Graduation came, and graduation went. I don't have too many memories of the day, except that Mum came over with my sister and it was nice to share the whole thing with them. It was never going to be a big deal for me, though, even before the disappointment of the result. I only went to university to get the qualification I believed I needed in order to have any chance of meaningful work in the humanitarian sector. So, in effect, I was merely going through the motions. In a desperate, final bid to move on after that one, final kick in the teeth, I burnt my graduation photos once they arrived a few days later. I still don't have any regrets. The closure felt good, and I looked silly in that hat, anyway.

Graduation day signalled the end of my university adventure and, despite the disappointment of the result, I was glad I'd at least done it and survived. It was, once again, time to take stock and reassess where I was, check out job opportunities and figure out my next move. Before that, though, there was the small matter of leaving my student house on Aberdeen Road and finding another place to live, this time with professional working people. Fortunately, a friend knew about a house about 200 metres up the street, on Upper Lewes Road, which had a spare room going for the summer. Within a week, I was in. There were seven bedrooms to go with a good-sized kitchen and lounge. It was a big house, and it felt as though there was a party going on in at least one of the rooms every night of the week. Living there, even just for a short while, was another adventure I hadn't planned. Life was nothing if not interesting.

Millennium

Given my record since arriving in Brighton three years earlier, it should come as no surprise to learn that the rest of that summer didn't go too well.

To earn a bit of much-needed cash, I returned to the language school where I'd worked in my first year and, once again, helped organise outdoor activities for the students. During an innocent kick-around at Preston Park one afternoon, my right knee suddenly jarred out of place and, for the second time in my life, the cartilage tore. It hurt. A lot. The first time I'd done this was during a five-a-side football match years earlier in Jersey, which led to my first operation. I'd forgotten how ridiculously, excruciatingly painful those tears could be – little else compares, believe me – and I had to be carried home by my colleagues. I stubbornly hobbled to work the following morning but could hardly move, my knee jammed at an angle. It was obvious to everyone, including me, that things didn't look good. Mobility is a fairly crucial requirement for an activities organiser, after all, and my summer was now effectively over. Within weeks I was in – and then, within a day, out of – the Princess Royal Hospital in Sussex, minus another chunk of cartilage. Although I'd be able to walk again within a couple of weeks, I couldn't run and my knee would remain highly unstable for the foreseeable future.

As I recovered, I turned my attention to finding more permanent accommodation and a more permanent job. Stability was the name of the game, for the time being at least. I found both fairly effortlessly, a colourful house on Blaker Street to live, and a job with Cable & Wireless to get me back into work. My previous computer experience secured the job for me, and an easy-going nature and some free-flowing banter with my future housemates secured the room. Within a few weeks Cable & Wireless had trained me up, given me a brand new, white Peugeot Expert van and assigned me to the Lancing office, a short drive along the south coast from Brighton.

At work I quickly made friends with Harry, a bald, edgy, rather intimidating-looking character with a great sense of humour and a heart of gold, whose main claim to fame was that he once appeared on the cover of *Brighton Rock*, a famous gangster novel (and, later, a film) based in the city. Within a few weeks we were both promoted to team leaders, managing around 20 fellow engineers. We were sent to work around the country, first to Manchester for a few weeks, and then to Surrey and other London locations. Digital cable television was new in the UK and we were all kept incredibly busy for months on end, earning as much overtime as we could and, later, helping Cable & Wireless hit challenging installation targets as part of a merger agreement with NTL. (That merger, which ended up eroding everything good that we'd built, came later.) For now, things were productive and fun and we had an exceptional team, the best in the country, as it turned out. We were the top-performing franchise that first year and were recognised across the UK for it.

Cable & Wireless took over my life for a while. If we weren't being loaned out to other franchises to help their often struggling teams, we were busy hitting our own targets. Without realising it, I'd drifted back into a way of life that I'd worked so hard to escape, one where work became everything, and you lived from payday to payday. On the plus side, I'd managed to get back into good shape financially, thanks to all the overtime I willingly worked. Saturday was our only real chance of a day off, if we weren't on shift, and Sundays were usually spent driving up to places like Manchester in preparation for the week ahead. It was a good five-hour drive and we'd all head up in a convoy, stopping along the way for coffee breaks and dinner. We were on double time on Sundays, so most of us easily earned a couple of hundred pounds for the drive alone. If work, and the money it brought, was all you lived for, then life was good. For a while the temptations were even too great for me, despite my ongoing struggles with purpose and meaning. Once again, I just fell into the same trap everyone else around me seemed to fall into.

With work so dominant I got little else done those last three months of the year. Before I knew it Christmas was upon us and, shortly after that, the dawn of the new millennium, an event which was predicted by some to signal the greatest technological breakdown in history. Of course, the so-called 'Millennium Bug' didn't destroy anything in the end. What also didn't happen was a party I had been invited to, somewhere in the Sussex countryside, to see in the year 2000. Plans abruptly changed and, just a couple of days before the big event, I found myself with nowhere to go to celebrate. All my friends were already off home with their friends and family. If I was to avert a millennium disaster, I needed to find something to do, and quick.

Each night, on my drive home from the Lancing depot, I'd pass an exit for Gatwick Airport. And each time I drove past I'd harbour strange thoughts of spontaneity, of randomly making the turn, abandoning my van in the car park, and jumping on a plane somewhere – anywhere – without telling anyone where I was going, or that I'd even gone. I still find the idea of vanishing fascinating, even if it's harder to do these days, and that it's totally unfair on family and other loved ones. But that aside, who doesn't ever feel like disappearing sometimes?

I may not have had the nerve to do the full disappearing act, but I did have dreams of showing up at the airport some day and just picking a flight somewhere. With my millennium plans in tatters, and quite literally no one else around, I knew there was no better time to try it. So I did. Passport in hand, I took a train to Gatwick with a sports bag packed with just a couple of changes of clothes, my music player and a little money.

Gatwick wasn't that busy when I arrived. It was around 2pm on millennium eve, so most sensible people were already settled into their hotels, or onto their second bottle of wine at a party somewhere. Only stragglers like me seemed to be left. As I wandered through the emptying departure hall, I looked up at the flight departure board. Most flights were heading too far away to make a short trip practical, but one, leaving in about an hour, caught my eye. Amsterdam. I'd been there before, so knew a little of what I was letting myself in for. And it was only about an hour away. I knew there would be plenty of bars, and likely plenty of other people, so why not?

I headed over to the KLM ticket desk and straight to the counter. There was no queue.
'Good afternoon. Can I help you?' asked the assistant.
'I hope so', I replied. 'Is there any space on the Amsterdam flight at 15.15?'
'Let me take a look. I know most of the earlier flights were fully booked.' Yes, I thought to myself. Most other people were definitely better organised than I was. Don't rub it in.
'It looks like we have a couple of seats, but a booking this late won't be cheap.'

I never could understand why airlines would want to charge inflated prices for seats that would otherwise go empty. Surely charging something was better than nothing? The flight was going to be a couple of hundred pounds, but when I considered the alternatives – and the fact there really weren't any – and all that overtime I'd been earning, I decided to throw caution to the wind. Sod it, I thought. When am I ever going to do this again? And it is the millennium, after all, a one-in-a-thousand-year event. I pulled out my passport and credit card, and handed them over.

Yes, I really was about to do this.

The flight passed without incident. I remember sitting in my window seat, alert with anticipation at what lay ahead. I was heavily into *Blur* at the time, as were many of my university friends, and played 'Coffee and TV', on repeat, pretty much the whole way. Once we'd landed I had nothing to collect – my sports bag had counted as hand luggage – so I headed straight for the train and the 15-minute journey into the centre of Amsterdam. Things were a lot busier than they had been at Gatwick, and there were hundreds of people like me who seemed to be heading into town for the millennium celebrations.

Unlike me, most of them had probably already arranged their accommodation, those planning on staying overnight, anyway. Despite having nothing booked, I was pretty relaxed about the whole thing. I'd not even considered trying to find somewhere to stay before getting on the plane. It wasn't until I'd trudged out of the first dozen hotels and guesthouses empty handed, all of them telling me they were already full to the brim, that I began to wonder whether my flippant, laid-back, casual approach to this trip was going to end in disaster. My return flight was in two days, and the idea of sleeping in a park for a couple of nights, however adventurous it may have sounded, wasn't appealing in the slightest.

For reasons I can't explain, I had a feeling that everything was going to work out, that someone was looking out for me, and that my bravery – or stupidity, perhaps – would pay off. There must be somewhere with a room, however shabby, where I could stay. As I made my way further and further down the street, I randomly took comfort in the story of Mary and Joseph. They eventually found somewhere, I thought. And so would I.

And, as it happens, I did. After an hour of searching there were literally no hotels or guesthouses left to try. As I became increasingly desperate I started to ask in the bars and pubs. There were plenty of those, and one must have a room tucked away somewhere, I figured. I pushed open the swinging doors of the next one I came to and, as I took my first step inside, the sound of rock music and the smell of stale beer mixed with marijuana instantly hit me. Well, this is Amsterdam, I thought to myself.

A few moments passed as I attempted to get someone's attention.

'Hi there. Do you speak English?' I asked the rather large, heavily bearded bartender as he wiped a glass with the stained cloth wrapped around his waist.

'Yes, I do. What would you like?' he asked in a thick Dutch accent, glass poised to pour me a beer.

'Oh, I'm not after a beer just yet. I was wondering if you had any rooms going?'

'Is it just you?'

'It is, yes. I'm looking for somewhere to stay for a couple of nights. Everywhere seems to be packed out.'

'Well, if you're not looking for anything too upmarket, we have a spare room in the attic of the bar here, right at the very top, which you could have. It is very basic though. We don't usually rent it out.'

I wasn't bothered about the condition, or the cost come to that. But, from memory, the price was pretty reasonable given my situation. I bit his hand off and took the room, quickly sliding my Euros across the bar before he could change his mind. It was a long way, up a few flights of dark, creaky, narrow stairs, and it was as basic as he'd said, so basic that I was once again reminded of where Mary and Joseph ended up. I threw my bag down on the single bed, pretty much the only piece of furniture in the gloomy, dull room, and headed back to the bar downstairs. Time to breathe, take stock, and finally start to enjoy this totally random, unplanned break of mine. Six hours earlier I had been in an empty house in Brighton and now here I was, in the middle of Amsterdam, about to have my first cold beer, just a few short hours before the world would welcome in the year 2000.

I've never minded my own company, and enjoyed sitting quietly by myself, drinking beer, contemplating, observing. I thought back to when I left Jersey, and about my time at university, about those wonderful three months in Uganda, and my work at

Cable & Wireless. It had been a hectic, often stressful and disruptive period in my life and, for the first time in a long time, I was now able to stop completely. I wasn't in the mood to make any effort to do anything at all, other than sit in that crowded, smoky bar and drink beer, alone with my thoughts. The lack of conversation was refreshing, relaxing, and a blessing.

As I moved from bar to bar on my very own one-man Amsterdam pub crawl, the inevitable happened and I ended up speaking to someone other than the barman (I had to order my drinks, right?). A small group of London bankers seemed to take pity on me and asked whether I wanted to join them. They bought me a drink and we all got talking. It was only a couple of hours until midnight, and I thought I'd probably done enough alone time for one day. It was time to put my sociable hat back on and rejoin the crowd in their growing excitement and anticipation. If all the computers in the world were going to explode in the next two hours, I might as well make the most of what little time I had left.

As midnight drew nearer we found ourselves edging ever closer towards Dam Square, a large, public space in the historic city centre where thousands of people were already bouncing around in full party mood, drinking and singing while they waited for the countdown and the blaze of fireworks. I don't remember too much about the countdown, to be honest – maybe one or two beers too many played their part – but in the morning I did have vivid memories of a firework exploding uncomfortably close to my left ear, probably a firecracker thrown by an overzealous party-goer. Reaching out to check my jacket, I realised I hadn't been dreaming after all. I could see straight away how lucky I had been. The left edge of the collar was burnt, small patches of melted, black, shiny blobs hung where there had once been a soft fleece lining.

On the second day I relaxed, drank coffee, looked in any shops that were open and generally took things easy. I went to my new, friendly neighbourhood bar for a couple of drinks that night and, this time, managed to keep to myself. I knew things were going to be as busy as ever at Cable & Wireless once we hit the new year, and was grateful for this one last chance to be somewhere neutral and alone to rest and think.

Once I was home, and the new year celebrations had faded into a distant memory, work did turn out as expected and I drifted back into the same old hectic routine. The one thing that made it all worthwhile was the great group of people I had around me. We continued to support each other when lesser teams would have cracked, and continued to hit our targets. Work was fun, enjoyable and we were still lapping up the overtime payments. These are the most dangerous times for anyone hoping to find purpose or meaning. It's easier to slip deep into a comfortable routine than to continue a difficult search, after all.

There were several benefits in working for a large company like Cable & Wireless, and for me comprehensive health insurance was one of them. A year on from my cartilage operation my knee continued to give me trouble, and simple tasks such as kneeling on the floor to wire up a digital receiver, or gently jogging for the bus, would cause considerable discomfort and pain. A few months into the new year I decided it might be worth calling them up. Would my health insurance cover my knee? I didn't hold out much hope – pre-existing conditions are rarely covered, after all – but there seemed no harm in trying. I had nothing to lose, so I picked up the phone.

A few weeks passed and I'd heard nothing, so I gave up on any hope of my knee getting fixed. Then, one morning as I was walking past my desk on a rare foray into the office, my phone rang. I very nearly left it, but doubled back to pick it up.

'Hello, Ken Banks speaking.'

'Oh, good morning Mr Banks. We've been trying to get hold of you for a couple of weeks now. It's great to finally speak. You made an enquiry about your health insurance, is that right?'

'Oh, yes. It is. And, yes, sorry to be so hard to track down. I'm rarely at my desk these days.'

'That's fine. We got you in the end! Well, the good news is it should be covered. We can send you a form to fill in, and then we'll make an appointment with a specialist in Haywards Heath. How does that sound?'

To say I was thrilled was an understatement. Five days later I'd filled in the form and sent it off, and shortly after I got an appointment. After an MRI scan, which would have set me back a few hundred pounds if I'd gone it alone, the problem became clear. I had a 50% tear of my anterior cruciate ligament (ACL), part of an intricate system of ligaments that pull together the two bones making up the knee. Mine was so compromised that the bones were able to move independently of each other, grinding on the cartilage that sits between them. The answer was an expensive ACL rebuild, and luckily I wouldn't have to foot the bill. It turned out that my consultant operated on professional footballers, so I knew I was in good hands. Within a matter of weeks I'd booked the time off work and was heading back to Haywards Heath on the train, sports bag in hand. The operation, where they grafted and screwed a piece of ligament from the back of my leg into my knee, was a great success. To this day I remain hugely grateful to the surgeon for his care, to Cable & Wireless for the policy, and to the circumstances that brought me into the office that morning when the phone rang. My knee had, and has, never been better.

After a brief period of rest and recuperation I was back at work, finally able to do the sorts of simple things everyone else took for granted. With a new knee and the kind of mobility I'd not had for years, things felt like they were on the up. Little did I, or any of my colleagues know, but clouds were gathering on the horizon.

A rival company, NTL, had agreed a merger with Cable & Wireless, and for the deal to go through we had to up our game and help the company meet almost impossible installation targets. Out of nowhere our team, once so dedicated, professional and supportive, began to struggle. Once steady, proud-of-their-work engineers found themselves having to cut corners, and quality suffered as a result. Morale, something that quickly dropped as our installation rate rose, wasn't something the new company seemed interested in. It was only a matter of time before people started to leave, and the best team in the country less than a year earlier slowly broke down, fell apart and disbanded.

Neither Harry nor I were happy, either. Maybe we'd had it too good, and it was only a matter of time before reality crept up and bit us on the backside. I'd never sought a career at Cable & Wireless, and had largely stumbled into the leadership role I found myself in. I'd had a good run, and got a new lease of life with my knee. That was priceless. Financially it had also been perfect timing after three years at Sussex University had made a considerable dent in my life savings. With a replenished bank account, my mobility back, a job on the wane and nothing to keep me in Brighton, I was once again free to pursue other opportunities.

I didn't know whether I was any closer to finding any kind of purpose or meaning in my life, but since leaving Jersey I'd explored all sorts of opportunities that, in one way or another, I hoped might help me find it. Sure, moving around constantly sounded new and exciting – adventurous, even – but the pressure to continually start over was becoming mentally exhausting and unsettling. Not only did I not know what I wanted to do with my life, but I increasingly didn't know where I belonged, either.

Despite the doubts, and my obvious lack of success, all that constant moving around was an approach I would stubbornly stick with. I had nothing to lose, and everything to gain, after all.

Next stop, Finland.

Suomi

I carefully scooped up as many meatballs as the large serving spoon would allow and placed them up against the pile of warm mashed potato already on my plate. After a pouring of gravy and a dollop of lingonberry jam, I paid the cashier and headed for a quiet corner of the staff canteen to eat. Despite the great food, I never did get used to those 11am lunches. In the midst of winter most Finns prefer to start work early, at around 7:30am, meaning they're free to leave while there's still a slither of daylight left to enjoy. As I was to discover, winter was tough, and people did whatever they could to make it as bearable as possible.

I finished up my early lunch, handed back my tray and headed up the metal stairs to the meeting room. My Business English students were already there when I arrived, punctual as ever. After smiles and greetings all around, I handed out the worksheets I'd prepared the night before, and another lesson for the reluctant teacher began.

My time in Finland was more of an accident than a smart, strategic move. I sold up and moved there in the autumn of 2000 to be with Elina, the Finnish girl I'd met a few years earlier while at university in Brighton. We became close over a four-year period, thanks to a mixture of emails, text messages and late-night phone calls, and both decided to give it a go. It was a standing joke among the Brits I met in Finland that we were only there because we'd fallen in love with a Finn. I'm happy to hold my hand up on that one, and many of the expat friends I made during my time there had similar stories to tell.

Despite having a stable team leader role at Cable & Wireless in the UK, I knew it was only a stop-gap in my pursuit of something meaningful. Moving to Finland seemed like an exciting, 'what have I got to lose?' kind of thing to do, something that might open my eyes to opportunities I'd not have otherwise seen. Cable & Wireless had been fun, and I'd been called to work across the country, helping roll out emerging digital television services with a brilliant group of engineers. Pat, my boss, had kindly held my job open for me, privately hoping my Finnish adventure would fail and I'd be back. He was to be disappointed.

This wasn't the first time I'd sold up and started all over again in a different country, of course, but it was the first time I'd moved to one with a language I could not read, write nor understand. I had learnt basic French and Spanish at school, and a little Italian during a brief spell dating a girl from Northern Italy, but Finnish was in a league of its own. I remember the street sign on the outskirts of Forssa, the town where I was to live, indicating the direction of the *aikuiskoulutuskeskus*, or the adult education centre to you and me. Finnish has a habit of throwing words together, not necessarily in the order you'd expect, to make long, almost unpronounceable ones.

Finnish has no connection with more familiar Latin or Germanic language groups, making it one of the most difficult languages for English speakers to learn. The average person would need to put in about a thousand hours to pick up enough to hold a basic conversation at the office water cooler. (Finland does have another official language, Swedish, which is a little easier to decipher, and most Finns speak it fluently.) Luckily for me, many Finns knew English to varying degrees, with younger ones keen to practise it at every available opportunity. Older people often had a basic understanding, but either a lack of confidence or a stubborn streak meant they rarely used it more than they had to.

I'd totally underestimated how hard it would be moving to a country with a vastly different culture and language to mine. The promise of new adventures can blinker you, I guess. When I first arrived I didn't have access to the Internet, and English newspapers and magazines were not only hard to find but expensive when I did. The limited number of TV channels were also void of anything I could make any sense of. Other than conversations with my partner and her family, or at parties with her younger friends, I was unable to communicate in any meaningful way and very quickly craved even the most basic of conversations with a native English speaker. In the beginning even popping to the shop, or getting on a bus, was fraught with danger as I prayed none of the locals would try to talk to me. I did eventually enrol with the local library, which had a small selection of English books, and took out a cable

TV subscription which gave me access to some international programming I could understand. Over time things did slowly get a little easier, but lacking something as basic as language can make you feel incredibly vulnerable. I can imagine how refugees feel when they're forced to start all over again in a completely new country.

My girlfriend lived in what was, for me at least, a dream wooden-fronted house nestled on the edge of a forest on the outskirts of Forssa, a nondescript town that most Finns had only heard of because they'd driven through it at some point in their lives. Reached by a dirt road and surrounded by a mixture of open fields and dense forest, her nearest neighbour was hundreds of metres away. It was calm and tranquil, and life seemed to run in slow motion whenever I was there. It was somewhere I could stop, think, read and contemplate life without feeling as though I was wasting time, even if I was. It was also the place where, four years later during a summer break, I would write the messaging system that would change my life.

The one-bedroom flat where I lived was about a ten-minute drive away in Forssa town centre. It was warm and comfortable with a small balcony facing the old church and town square. The town centre didn't have a huge amount to offer, just a couple of supermarkets, a pizza restaurant, a kebab house, the odd cafe and bar, a cinema, and an occasional open-air market. Everything was within walking distance, and I'd spend my free time strolling around the often deserted streets, hunting out English books and newspapers or discounted microwave dinners in the supermarkets. Buses passed by regularly on their way to St Petersburg, a reminder that the Russian border was close by, something I found rather unnerving as someone growing up, albeit at a distance, during the Cold War.

For the Finns themselves, the Cold War had been anything but distant. Finland shares its Eastern border with Russia, after all, and they've had to live under the constant threat of attack or invasion for longer than many older Finns care to remember. Their country's worst fears came true on 30th November 1939 when, after failing to wrestle territory away from the Finns diplomatically, the Soviets took the decision to invade 'on grounds of national security'. This so-called Winter War broke out just three months after the German invasion of Poland, an event that signalled the start of the Second World War.

Despite their overwhelming strength and firepower, the Russians toiled for three months against the determined but ill-equipped Finns, suffering severe losses and making little ground in the freezing conditions. After a decision to pause, regroup, and deploy new tactics, the Russians tried again a month later and fared better, although they still suffered serious losses. The Finns finally ran out of steam and gave up the fight in March 1940, leading to an interim peace. The end of the Winter War wasn't strictly the end of hostilities, though. The two sides went on to fight the so-called Continuation War until 1944. After the signing of the Moscow Peace Treaty the Soviets ended up with 11% of Finland's territory as reward for their efforts, territory they still hold to this day. In exchange, the Finns retained their independence. Known as the 'Karelian Question', the Finns still talk about whether they should try and get some of that territory back, 80 years on.

The Finns are a very proud people, and the way they stood up for themselves all those years ago remains a great source of national pride. Many put it down to something they call *sisu*, a word which has no direct translation but describes their inner strength, perseverance and determination, often against all odds – precisely, in fact, the scenario they faced in the Winter War. *Sisu* has been described as everything from the social glue that defines Finland and its people, to a 'second wind', something you get when you've got nothing left in the tank and all else has failed. I certainly knew that feeling well.

Sisu was just one of the reasons Finland bore little resemblance to any other country I'd been to. Sure, there's the weather, and the endless summer days, but there's also a tremendous feeling of space, no surprise given it's the third most sparsely populated country in Europe with an average population density of just 19 people per square kilometre. Wherever you drive or walk, you can literally feel the emptiness around you. Many-a-time I'd notice, as I drove between offices for my English lessons, that mine was the only car on the road. Instead of people you'll find an incredible 188,000 lakes, and anywhere you don't find a lake you'll almost certainly find trees. Three-quarters of Finland's total land area is forest. The air always feels wonderfully fresh and alive, and everybody and everything seems intimately in tune with the natural world around them. Regular trips into the forest felt like a walk through a childhood fairy tale. Sometimes we'd go blueberry picking, sometimes for beers by the edge of a lake, or sometimes for an impromptu barbeque or swim. Only the occasional (but relentless) swarm of mosquitoes ruined the experience. They seemed to have a particular liking for English blood, and were more vicious than anything I'd ever experienced, even during my trips to Africa.

Wild swimming wasn't something I'd tried much before, but I made up for it that summer. I'd been an avid swimmer after finally taking the plunge and learning in my late twenties after my brush with death on the Zambezi. The beauty of swimming in a Finnish lake, other than being surrounded by dense forest under a bright, deep blue sky, is that you're usually the only person around for miles, and you're pretty much free to go anywhere you like. Swimming in the many public pools, on the other hand, is a different experience altogether. About 20 minutes' walk from my flat, on the far edge of town, was the local sports centre. Along with a pool it was equipped with a good gym and badminton courts, and a small café. As someone who strived to keep fit and healthy I'd visit when I could, either for a swim or to lift some weights. Keeping fit seemed to give me the balance of mental and physical activity I needed as I lived my daily routine, while continually striving to move my life forward.

If you randomly find yourself in Finland, and you fancy a swim in a public pool, a word of warning. Swimming shorts, or any kind of shorts, are banned, and the Finns are sticklers for rules. Speedos are the order of the day. At first I was blissfully unaware of this, so I remember my first swim well for all the wrong reasons. I'd just got changed and, after washing my feet in that little shallow area you always find poolside, I strode purposefully towards the steps leading into the water.

     'Anteksi! Anteksi!'

I turned around. Was this guy shouting at me? The lifeguard seemed to be looking in my direction.

     'Sorry, do you mean me? Do you speak English?' I asked, stopping in my tracks.

     'Joo, joo. Sorry, I mean yes. I do.'

     'Oh, okay. What is it?' Taking a few steps closer, he looked and pointed down.

     'Shorts. Shorts are not allowed in the pool. Swimming trunks only. Sorry.'

I was allowed to swim that day, but on condition I wore proper trunks next time. That didn't stop me getting my fair share of evil looks from other swimmers, mind you, who seemed disgusted with my blatant disregard of the rules. It's fair to say, I've had better days at the pool.

Not content with forcing people to wear unflattering swimwear, many pools also boast small 'dipping pools' which you're welcome to climb down and briefly immerse yourself into. That in itself might not sound particularly adventurous, but the water is a chilly ten degrees Celsius, which is. The Finns have a thing about throwing themselves into freezing cold water, sometimes straight from a steaming hot sauna. It was something I'd often try, and an experience I would never forget.

Public saunas were another thing I'll never forget. The Finns may be better known for their rather quiet, introverted nature, but they're not afraid to bare all in the sauna. Always one to immerse myself in the local culture, I braved it many-a-time after my swim, joining a dozen or so naked, often bald, plumpy, older men on the wooden benches dotted around the sides. One person, usually the closest, took responsibility for managing the temperature by throwing a small ladle of water onto the hot coals whenever it was deemed to have dropped too much. Occasionally, just for a laugh – or perhaps to wind up the English bloke – they'd throw extra water on for no obvious reason. I was always amazed at how quickly the temperature rose. You could literally see the cloud of hot steam gather pace as it thickened and drifted across the tiny space, penetrating deep into your bones as it wrapped itself around you. It seemed to be a matter of male pride to see how much you could take. Eventually it would fill the room, my eyes steaming over, reminding me of the inside of a car windscreen on a cold, wet winter's day. During the summer months I would relish the opportunity to run straight from a hot sauna and dive into one of the many freshwater lakes.

Everything is always new and exciting when you first arrive in a country, so there was plenty to keep me occupied. But I did need to find work, and did need to pay my way if I was to avoid a premature return to Pat and my job at Cable & Wireless. Fortunately, it wasn't long before I fell into the world of Business English. If I'm honest it was the only thing I could realistically offer, despite having no teaching qualifications, but I was there and could start immediately. For the English schools that decided to hire me, that turned out to be more than enough.

I had expected my inability to read, speak or understand Finnish to be a major handicap as a teacher. You'd think knowledge of the local language would be important but, as it turned out, not being able to say much beyond the usual 'hello', or 'how are you?' or 'my name is Ken' worked a treat. Knowing that I wouldn't be able to switch to their native tongue when things got difficult, my pupils had little choice but to try and figure things out in English, however difficult they found it. Sure, sometimes we reverted to miming, doodling or reaching for props, but it always made for a lighter moment and, on the whole, it worked well. I'm convinced many of my students thought, deep down, that I was fluent in Finnish and was just being crafty feigning my ignorance. I wish I'd been that clever.

Despite the challenges, my fleeting teaching career was both enjoyable and surprisingly successful. Class attendance remained high, companies kept asking me to come back, and executives whom I taught, one-to-one, continued to book lessons. I relished the freedom that teaching conversational Business English gave me. It was less about the grammar and the finer points of the language, and more about how my students might hold their own in a conversation at a conference, or the different ways they might speak to someone on the phone. I remember a series of lessons where we did nothing but focus on small talk, something many Finns are surprisingly bad at. It was never lost on me how strange it was having to reassure fully grown adults that it was okay to talk to people you've just met about the weather, or their favourite football team, or where they're from, or about their family or their favourite food, even their names. The Finns seem to be genetically programmed to get straight to the point, something which has its time and place, for sure, but not at the bar after eight hours trapped in a stuffy conference room.

Over the course of my three terms I was called on to teach escalator engineers, chemists, drug manufacturers, margarine producers, university lecturers and CEOs how to talk about the weather – a subject which, as a Brit, I happened to be something of an expert in. I made many good friends, and we all looked forward to spending our time together. To this day, I don't know how much my students actually learnt, or how much better their English became, but for many it was fun to escape their desks for an hour and learn a little more about a language spoken by over 1.5 billion other people on the planet. And for me, getting paid to basically talk was a dream job.

I had use of the company car, a small black Ford Ka, to help me get to some of the more remote lessons. While some of our clients were close by in Turku city centre, others were a bus ride away, or out in the sticks. I used to enjoy the journeys, eventually getting used to driving on what, for me, was the wrong side of the road, and I took the opportunity to relax, see some of the beautiful countryside and listen to Finnish pop music. I'd discovered a few bands that I liked, and enjoyed listening to the sounds of the words, words I had little understanding of, of course. When I heard music that I liked on the radio it was always fun trying to figure out what the name of the song might be. I often drew on my students to help with this – one of the perks of the job, I guess.

Time passed by surprisingly quickly. Before I knew it my third term was over and I was on a bus heading home from Turku for the last time. There was no need for the company car anymore. Stepping into my second-floor flat, I gently closed the door behind me, dropping my bag down on the table. Taking a deep breath, I collapsed on the sofa. It had been a busy few months, and my final Business English class of the summer was done and dusted. My students, along with pretty much every other Finn I knew, were now slipping into holiday mode, abandoning their offices and fanning their way out into the forested countryside. Barbecued sausages, beer, fishing and saunas, all in the comfort of quaint, lakeside summer cottages, was the prize awaiting many of them. It was mid-June and little stood between day or night, summer becoming everlasting and magical. I never did get used to waking up and thinking I'd overslept, only to find it was still the middle of the night. Summer presented a brief opportunity to do a little living in a year dominated by the dark and cold. Finns lived for this time of year and, with almost everything now closed, it showed.

'Have a great summer, Ken. Hopefully we'll see you in September', said Mike, one of the directors at Kielikanava, the language school where I did most of my teaching.

'Thanks, Mike. You, too. I'm looking forward to having a break and just taking some time out. I need to decide what to do next.'

Mike was originally from Manchester and had moved to Finland many years earlier. He was a cheery character with a neat beard, and was great company. I admired his command of Finnish and felt wholly inadequate in his presence. We'd become good friends despite supporting rival football teams back home (he was, naturally, a Manchester United supporter, and I'd been a Liverpool fan from a very early age). Little did I know, but that was the last time I was to see Mike until I returned to Finland on a working trip four years later.

I was now officially on holiday, although unemployed might be a better way of describing it. My boss, a wonderfully kind and gentle lady by the name of Riita, was happy and grateful that I'd taken on every piece of work she had offered me. She later told me that, financially, it was one of the best years they'd had in a while. Teaching was a bit of a no-brainer for me and, as a people person, I got a kick out of planning lessons and having the freedom to see where all those conversations went.

Despite having an enjoyable nine months immersing myself in a totally different culture, while experiencing weather that went from -25 degrees C and deep snow to +35 degrees C and blistering sunshine, living and working in Finland was only ever likely to be a stop-gap for me. Five years had now passed since I left Jersey and I was now in my mid-thirties. Although I'd done a lot in that time, I don't think I was any closer to finding out what I wanted to do with my life. If life really was about the journey and not the destination, I was doing pretty well, but I was desperate to know what, if anything, lay at the end. With the English schools closing I was now out of work again, and drifting. I could wait until autumn and restart my teaching when the Finns returned from their extended vacations, or I could leave and look for answers elsewhere. Other people might have taken the summer off and just relaxed – sausages, beer, sauna, sunshine and lakes were tempting, of course – but that wasn't really an option for someone like me. With the clock ticking, I was becoming increasingly restless.

Despite the relative comfort and ease of life in Finland, I was still anxious to continue the search I'd started all those years earlier. I'd not been long out of Sussex University and was still hungry for knowledge about the world around me. If I had time to spare, and no obvious plans, I could do much worse than return to studying, I thought. I had an old, borrowed computer in my flat, with dial-up Internet, and decided it might be worth exploring academic opportunities in Finland and beyond. I had nothing to lose, after all.

After a couple of days I ended up with a shortlist of three – a Masters Programme in primate conservation at Oxford Brookes University in the UK, indigenous studies and anthropology in Canada, and international development (with geography) in Finland. I made enquiries about all three. Oxford Brookes offered me a place, but I'd need funding and going back to the UK felt like a step in the wrong direction. Canada was a little trickier given visas, travel and the cost of yet another relocation, but it was a country I'd always dreamed of visiting. The Finland option made most sense though and, for once, I decided to pursue the path of least resistance.

I phoned the University's Admissions Department.

'I've been looking at your Masters Programme in international development and geography, and wanted to check entry requirements and any costs', I said, explaining that I already had a bachelor's degree from a UK university, but couldn't speak or write in Finnish.

'Let me see. Well, a good undergraduate degree would be a requirement, and it sounds like you have that. And the language of instruction is English, so that should work fine', said the lady, in almost perfect English, on the other end of the line.

'So, how much does it cost?' I asked. My funds were tight, to say the least.

'Are you a British citizen?'

'Yes, I am.'

'In that case, you'd qualify for the Programme for free.'

'Free?'

'Yes, free. So, it probably is worth considering an application. Would you like me to send something out to you in the post?'

'That would be brilliant. Thank you!'

After sharing my address I put down the phone, unable to believe my luck. It all seemed too good to be true.

A few days later, after digging into the Programme a little further, I decided to take a bus to Helsinki to get a better sense of the place. The masters course was a natural extension of my earlier studies, and ran for one year. The tutors were interested in my earlier work and felt my undergraduate degree was a solid foundation. Despite this, they still wanted me to take an entrance exam. The Finns take education very seriously (they're not top spot in the world for nothing). Their caution was understandable, I guess, given I'd effectively just walked in off the street.

An entrance exam, in a subject I was passionate about, in English, seemed eminently achievable. Sadly, as with most things in my life up to that point, things didn't turn out to be as straightforward as that.

96

A few days later I received a letter giving me information about the exam. I was genuinely excited at the opportunity and was already looking forward to sitting it. Then, as I started reading, it hit me. Flicking through the paperwork I noticed that the reading list was in Finnish, and all the books on it were Finnish. If that wasn't enough, the exam would also be in Finnish. Let me get this straight, I thought to myself. The Masters Programme itself is all in English, but to prove I'm up to the job I need to sit an exam in a different language, and one I can neither speak nor write. I slumped back in my chair, my head in my hands.

I have to admit I shed a few tears of frustration that night. I didn't want to be an English teacher, and with my academic dreams seemingly up in smoke I was back to wondering what I could do next. That night I burnt the University's letter on my balcony. Looking back it wasn't the smartest thing I've ever done, but it gave me a sense of closure, not to mention a twisted sense of revenge, and it was better than throwing something through a window. That night another door had been slammed shut in my face. I went to bed feeling depressed and as lost as ever.

The next day I did my best to pick myself up and got straight back onto the computer. I decided to build a website, my first, to share my interests, hopes and dreams with the wider world. Somebody out there must be looking for someone like me, I thought. I had no idea what I was really doing, and 2001 was still relatively early for the World Wide Web, but I registered igisi-hill.com – named after one of my favourite camp locations on the Uganda trip I took in 1998 – and started crafting my call to the world. Looking back, it was more a cry for help, to be honest.

I grabbed a coffee, sat down at my desk, composed myself and started to type. Emotions were still running high after the University letter, which was now reduced to a pile of ashes in the bin.

My first ever website started with these words:

'This website has appeared purely and simply due to a lack of results in my search for academic study, research opportunities or what I would call 'meaningful employment'. What it certainly is not is a celebration of my achievements to date. It is self-promoting, not self-glorifying.'

What followed turned into a mixture of rant, manifesto and plea for divine intervention. I laid bare my hopes, dreams and aspirations, and my frustration at not being able to find an avenue for my passion and energy. My website ended with a call for help, for someone to give me the chance I was desperately seeking.

'Whoever and wherever you are, hopefully you'll be in a position to take me along on your journey, or know someone else who may. Either now, or in the very near future. Thank you.'

So there you have it. My Finnish adventure ended with a desperate cry for help to no one in particular on a website nobody knew was there. Not surprisingly, and not for the first time, my efforts went unrewarded. The universe seemed to be telling me something. Failure was becoming a habit, something I didn't like one bit. Giving up, though, never felt like an option.

Weeks later I made the difficult decision to sell up and return to the UK.

Back to square one.

Primates

It doesn't matter how many times you've been to Africa, nothing quite prepares you for Lagos. Admittedly I've never been one for big cities, not even at home in the UK. I prefer the rush of fresh air and exhilaration of open countryside. So arriving in Africa's largest city that warm December evening in 2001 was never going to be easy. As I stepped off the plane at Murtala Muhammed International Airport I was immediately struck by the heat, and a multitude of sounds and smells that can easily overwhelm the senses of even the most hardened traveller. Ground staff ushered us towards the twin doors of the terminal building and, after collecting my rucksack from a rickety, squeaking belt, I left the bright lights of the arrivals hall behind me and headed into the chaos outside. Hawkers peddled their wares left, right and centre, attempting to sell me everything from hotel rides to mobile phone SIM cards and watches. I gazed through all of them, nervously scanning the crowd for Zena, the founder of the sanctuary where I was to work.

Lagos is not a place for the faint-hearted. My only comfort was knowing I'd only be there for a couple of days before heading off to calmer climes in Cross River, a state which boasts the last remaining tract of rainforest in the country and, yes, a river (two, in fact). I couldn't wait to set foot in the place that would become my home for the next year.

I've never been a nighttime flyer, preferring to arrive somewhere in daylight hours allowing me time to get a sense of a place, usually on foot, before the onset of darkness. Of course you don't always have a choice, and this was the only flight British Airways operated into Lagos. Still, I did manage a faint glimpse of the airport out of my window as we approached, enough to see rows of Nigerian Airways aircraft parked up away from the main terminal. That year the UK had decided to halt direct flights by Nigerian Airways between London and Lagos, citing safety concerns, and the airline was in decline. Within two years they would only have one serviceable aircraft left, an ageing Boeing 737. They eventually ceased operations in 2003 and were liquidated shortly after. Despite my discomfort at night flying, I was acutely aware that things could have been worse and that, had it not been for the ban, I could have been on one of those older planes. Count your blessings, I thought.

I don't remember much about the seven-hour flight over from London, although I do recall being a little more apprehensive than usual. I was travelling alone and was about to meet people I hadn't met before, in a place I'd not been to before, to do a job I'd never done before. It's probably no surprise but Nigeria never did make my 'countries I'd love to visit' list, and recent troubles in the Niger Delta, where I was to spend most of my time, did little to reassure me. Protests against oil giant Shell – who were accused of indiscriminate pollution, corruption and the support of military crackdowns against civilians in Ogoniland – had turned violent, resulting in the 1995 hanging of nine Nigerian activists, including writer and environmental campaigner Ken Saro-Wiwa. The international outcry did little for Nigeria's already battered reputation. To the outsider, the country seemed brutal. It probably felt that way to many of the people living there, too. Having an army run a country often does that.

I'd taken a giant leap of faith heading to Lagos. A couple of months earlier, following my unexpected return from Finland, I'd volunteered to manage a primate rehabilitation centre in Calabar, southern Nigeria after randomly finding a vacancy online. I sealed the deal with Bob, one of the charity's trustees, over a steaming hot latte and an almond croissant in the chilly, echoing concourse of Brighton railway station. Bob wanted to make sure I knew what I was letting myself in for.

'Nigeria can be a pretty brutal place', he told me.

'From what I've heard and read, I get that, Bob', I replied, trying to reassure him.

'It can be hard work getting even the simplest things done, and you'll be living very simply with few, if any, of life's luxuries. Don't expect it to be fun, although it will be very rewarding.'

There was little chance Bob was going to put me off, but to convince him that I'd experienced similar conditions before, I showed him my scrapbook from the conservation expedition I'd been part of in Uganda in the summer of 1998. Photos of us camping out in the bush for three months did the trick.

Primates had long fascinated me, and I still have vague childhood memories of visits to Jersey Zoo, where the keepers would often mingle among the visitors with baby gorillas wrapped around their shoulders. (That kind of behaviour is now frowned upon, of course.) My later work at the zoo, updating their computer systems, meant I had ample opportunity to walk the expansive grounds at will, often at night, making friends with the staff and getting to see things behind the scenes. I remember wanting to be a keeper as a child, an ambition that lay somewhere between wanting to be a train driver or a scientist. Life, however, took me in a different direction. I almost veered back towards a professional conservation career in 2000 after being offered that place at Oxford Brookes University, but I decided to get some practical experience instead. Once again, my instinct to see and do things for myself won over.

Fortunately, the management and technical skills I'd picked up during my time in banking and information technology turned out to be in high demand in the conservation world, giving me something of an easy way in. Volunteer keepers were relatively easy to find. Volunteers with business skills were not. Being one of a few willing to work in Nigeria also gave me a clear advantage. Many people I knew didn't want to go anywhere near the place, preferring the gentler pace of life on the other side of the continent.

Cercopan, the sanctuary where I was to work, focused primarily on long-tailed forest monkeys such as guenons and mangabeys. The more romantic side of primate conservation, chimpanzees and mandrills (those with the brightly coloured faces and menacing teeth, made famous in *The Lion King*) were the speciality of Pandrillus, another local sanctuary run by Peter and Liza, an unlikely conservation duo. Despite being just a short drive across town, it was a journey I'd make surprisingly few times. Conservation can be a competitive business, as it turns out.

Few of the people I was to work with that year had any experience of life outside a dictatorship. The country had been under military rule since 1966, interrupted only by a brief period of democracy between 1979 and 1983. It wasn't until 1999, following the peaceful transfer of power to President Olusegun Obasanjo, that they experienced their first proper taste of political freedom. The transition was a challenging one, though, and the economy was still in tatters when I arrived two years later. Corruption was rife, with Nigeria one from bottom in Transparency International's 'Corruption Perceptions Index'. Only Bangladesh fared worse. I'd experienced corruption at civilian level during earlier travels in Zambia and Uganda, but once you learnt to recognise it, and knew how to deal with it, it was relatively easy to navigate. Having guns waved in your face by irate Nigerian police or soldiers on the other hand, as they demanded money, was something I never really got used to.

Friends and family didn't expect me to last long in Nigeria, if I'm honest, and many expected me to be home within a few weeks, either shaken up but in one piece or, most likely, in a box. They were nearly right about the box, although I did last much longer than a few weeks.

My first night in Lagos was spent on Zena's couch. The couch, along with the humidity, unfamiliar sounds and smells and general apprehension all combined to make a good night's sleep somewhat elusive. I always found myself torn in situations like this, knowing that the comfort I enjoyed was totally at odds with the sacrifices I knew I had to make if I was to find some kind of meaning and purpose. As a child, simply getting the bus to town on my own was a big deal and now, here I was, alone and thousands of miles from home, in a country that few people I knew had anything good to say about. Somewhere out here was my meaning, my purpose in life – or so I believed – and I thought that getting out of my comfort zone was the only way I was going to find it.

Was it going to be in Nigeria, though? Suffering, whatever physical or mental form that took, was the price I decided I had to pay in order to find out. It felt like a test of my commitment, something I had to do to prove I was serious about finding my way in the world. 'No pain, no gain' was a mantra that often sprung to mind during my time there.

I wasn't the only arrival that December evening. A baby chimpanzee had been confiscated from a local market a couple of days earlier and was waiting to be collected from Lekki, a conservation and education centre in Lagos run by the Nigerian Conservation Foundation. Primate rescue was to be a theme of my time there, as was a sense that a large part of the conservation effort was really just a damage limitation exercise. Rehabilitating orphaned primates was often the easier part, even though it was hugely challenging and distressing. Changing perceptions, overcoming local politics, protecting forests and trying to shift cultural mindsets turned out to be much harder. Most of these things can take an incredibly long time, time that many species in countries like Nigeria simply don't have.

My first breakfast was overshadowed by the anticipation of that first rescue. I'd never been anywhere near a primate before and never expected to get the chance so early in my trip. Zena and Sarah, another colleague I'd met the night before, took charge. Between them they had done this dozens of times, a record no one was particularly proud of, and evidence of the extent to which rampant bushmeat hunting and wildlife trafficking took place across the country. The fact this chimp was being sold openly, in a food market in the middle of the capital city, showed just how ineffective wildlife conservation laws were, and how seriously people took them. All that aside, this was to be a relatively straightforward rescue, a young chimp orphaned and put up for sale as a pet after its mother was likely killed for food. Lekki's rangers had done the harder work getting the chimp out of the market. This time around, our job was merely to collect it and transport it to Peter and Liza at Pandrillus in Calabar. Some of the later rescues I was involved in were far more difficult, and far more heartbreaking than this. The first rule of primate rehabilitation is to not get too emotionally close. Given my overly sensitive nature, that was always going to be more than a little difficult for me.

So my second, and final, night in Lagos was shared with a baby chimpanzee which we'd made as comfortable as possible in a large, wooden crate, complete with a soft blue beach towel which, in places, poked out awkwardly between the planks. I hardly heard a peep from him that night. Primates this young rarely leave their mother's side, and it was impossible to imagine the trauma it had gone through to get here. I was later to witness monkeys coming in with broken limbs, bones snapped to loosen the grip around their dead mothers. This chimp was one of the lucky ones, if that's ever the right word to use in situations like this. The more time you get to spend with primates, the more human you realise they are.

Primates come in all shapes and sizes, with each species having evolved in its own unique way, often influenced by their environment. When most people think of primates they think of gorillas, chimpanzees or orangutans, but lemurs, bushbabies and lorises are also all primates, as are humans. The term 'primates' captures the whole range, but monkeys and apes are quite different from each other. Monkeys have tails and apes do not, and apes are considered more intelligent. Monkeys native to Africa and Asia are often referred to as Old World monkeys, but about 30 million years ago isolated groups somehow drifted their way to Central and South America, and Mexico, giving rise to New World monkeys. The vast majority of primate species, but not all, are threatened in some way due to habitat destruction, usually for large scale industrial farming, or poaching.

Chimpanzees or chimps, like the one we were transporting to Calabar, have coarse black hair over most of their bodies but none on their faces, hands or soles of their feet. Even at a young age they can be incredibly strong and need careful handling. Hunters usually tie them up, compounding their pain and anguish yet further. Today there are fewer than 250,000 chimpanzees remaining in the wild. They have already disappeared from four African countries, with poaching and hunting for bushmeat taking a terrible toll on their numbers.

The following morning Zena, Sarah, the chimp and I all headed to the domestic terminal at Murtala Muhammed International Airport to catch our flight to Calabar. The airport was as chaotic as usual, and I was already looking forward to turning my back on Lagos. Between me and my final destination was a 75-minute flight on Chanchangi Airlines, a small domestic airline with just a handful of Boeing 737's. Flying internally in a country infamous for government corruption and a poor culture of maintenance had always been one of my biggest concerns, and the three Corporate Merit Awards taped to their check-in desk for 'Best Domestic Airline of the Year' did little to reassure me. Check-in wasn't much comfort, either, with no obvious structure to the process and random people seemingly able to get boarding cards with no obvious sign of a ticket. As we sat there for hours, waiting for our increasingly delayed flight (no one seemed to know where the plane was for most of the day) my nerves simply worsened in the searing heat and humidity.

Our plane did eventually arrive in the early evening, something which seemed increasingly unlikely as the day wore on, causing a stampede as the check-in area rapidly emptied and a swarm of passengers raced towards the bottom step of the aircraft. There was no queue to aim for, just, quite literally, the bottom step. That was the prized spot for anyone hoping to guarantee their seat on the plane. Many of them had done this before, and it showed.

It wouldn't be long before I realised why. The check-in staff had issued about 125 boarding passes for a flight with closer to a hundred seats. I was one of the lucky ones to reach the bottom step in reasonably good time. I edged my way to the top, turned into the aisle and began to weave my way past groups of arguing passengers to my seat, which was fortunately still vacant – all thanks to Zena and Sarah who continually pushed me forward through the mele. If this was the kind of service provided by 'Domestic Airline of the Year' then I couldn't help but wonder what it might be like to fly with the runner-up.

I made myself comfortable, trying to hold my nerve amid flying elbows and the ensuing madness, while wiping my brow in increasingly futile efforts to keep cool in a stuffy cabin, jam-packed with way more people than it was ever designed to hold. Just when I thought I'd somehow made it, out of the crowd a passenger with a menacing look on his face caught my eye, heading purposefully towards me, waving a creased up boarding card.

'You have my seat!' he exclaimed, pointing to the seat number on his boarding card with one hand, while glancing up at the seat number above my head. 'I want my seat', he demanded.

'My boarding card also has this seat number on it', I replied. 'I'm not sure what's happened here.' Of course, I did know what had happened, but I still wasn't going to move.

One of the flight attendants spotted our exchange and quickly worked her way over. I was already under strict instructions from Zena not to move under any circumstances if I was challenged, and the attendant seemed to agree that it was my seat simply on the basis that I'd got there first.

As they argued I stayed put, not knowing whether I'd ever make it to Calabar if I gave up my spot. After some lengthy discussion, most of which took place at high volume and in Pidgin English, the stewardess managed to convince my opponent to try for a later flight, whenever that would be. As the aisle slowly emptied and the lucky passengers settled into their seats, I finally began to relax, knowing I had cleared the first hurdle of my first Nigerian domestic flight. Now it was just the simple matter of taking off, cruising and landing in one piece.

As we waited for the doors to close I began an unconscious, nervous inspection of the interior of the aircraft. I shouldn't have. While the plane appeared to be in better shape than I'd expected, the reason soon became clear. This wasn't one aircraft, but the best bits of lots of aircraft, all stuck together. The 'No Smoking' light was in Spanish, the 'Fasten Seatbelt' indicator in French, and the 'Life Vest is Under Your Seat' sign in a language I didn't recognise. Before I had time to look any further, we were accelerating down the runway and into the air, the wheels gently thudding as they retracted into their bays under the wings. There was no going back now. At least we'd taken off safely. Just another 74 minutes to go.

Thanks to the extended delay our day flight had turned into a night one, compounding my nerves yet further, and I couldn't see anything out of the windows except for the occasional faint flicker of gas flaring from rigs off the coast. We were served refreshments, a sickly sweet fruit juice and a bag of peanuts that most of the other passengers couldn't seem to get enough of. Other than that, the flight passed without incident, and there was little turbulence until we neared Calabar. Hot air is generally thinner than colder air, meaning aircraft often struggle to get the lift needed to take off and land in hotter countries. As a result we hit the ground harder and faster than usual, but at least we'd landed.

Margaret Ekpo International Airport, otherwise known as Calabar Airport, was a far cry from what I'd experienced in Lagos. Sure, it was getting late and we appeared to be the only flight to have arrived within the last hour or two, but you weren't jostled as you walked through the terminal, there was no scrum of passengers around the baggage carousel, and no sales people waiting to jump you as you left the building. The atmosphere just felt calmer. I was grateful to have arrived in one piece, and took comfort in the knowledge that I'd not have to set foot in another airport again, or another cobbled together aircraft, for at least twelve months. Or so I thought.

I was finally in Calabar, and my conservation work could begin.

Nigeria is a huge country that boasts a variety of different habitats ranging from mangrove swamps and rainforest to open savannah. The Niger Delta is home to most of what remains of Nigeria's once significant rainforest belt, which once stretched across the entire southern part of the country. Forest loss, particularly old-growth forest, has accelerated over recent years due to the pressures of a booming population, demand for agricultural land, mining, and both legal and illegal logging. Things were particularly bad when I arrived in 2001. Between 1990 and 2005 the country lost a massive 80% of its old-growth forests. Many of the hundreds of species of birds and mammals, dozens of species of reptiles and amphibians, and numerous species of primates, continue to find themselves under direct threat as a result. Larger primates, such as the Cross River Gorilla (which only number a couple of hundred) and chimpanzees and drills are particular favourites of hunters and wildlife smugglers. But even the smaller, long-tailed forest guenons, such as the Sclater's, putty-nosed, red-eared and mona guenons, are also targeted. These were the primates that Cercopan, the sanctuary where I was to work, had decided to focus its efforts.

Cercopan was founded in 1995 by Zena Tooze, a Canadian conservationist with a background in veterinary nursing and zoology, after witnessing first-hand the plight of primates in Nigeria. In 1997, Cercopan held its first environmental rally in Calabar, going on to employ its first education officer a year later. Education was to be an ongoing theme for the sanctuary, with rehabilitated primates in the Calabar headquarters primarily acting as 'ambassadors' and a main focal point for local education efforts and awareness-raising. In 2000, Cercopan signed an agreement with the community at Iko Esai, a village roughly three hours' drive and hike away, to allow the building of a research camp in what remained of their community forest (rehabilitated primates were eventually released into large, open-topped enclosures in Rhoko Camp, as it was known, a year after I left, in 2003). By the time I'd got there Zena was spending increasing amounts of time with her husband in Lagos, so the sanctuary needed someone on-site to run it. I was to be next in line of a growing number of volunteers to take up the challenge. Suzanne had done her stint and was about to head back home to Canada.

The continuous cycle of volunteer managers might have presented great opportunities for those of us coming in, but I felt sorry for the keepers who were often at the rougher edge of the change. They must have had so many questions. Who would this new person be? What would they be like? Would they go easy on us? By the time they'd got comfortable with one manager and his or her preferred routine, someone else would come in and things would change again. I decided to call a meeting and reassure them early on.

One day, after the morning routine was complete, we all gathered in a circle under the rattan roof of the education centre. Bassey and Wayas, our nighttime security, stayed back an extra half-hour so they could join us.

'Hi everyone. It's really great to be here with you. My name's Ken and I'll be working with you all for the next year, taking over from Suzanne when she leaves in a few days. How are we all doing?' Everyone looked understandably nervous. The purpose of the meeting was to reassure them as best I could. They all responded with welcoming but awkward smiles. I continued.

'Let's go around and do short introductions, and then we can talk a bit more about what I hope we'll all be able to do here.' One by one, they went around the circle, telling me their names – Pius, Austin, Abakum, Glory, Egu, Wayas, Eme, Debo, Bassey.

'Thanks, everyone. It's nice to meet you. Now, I know it's difficult with people coming and going all the time, but I wanted you to know that I'm just a visitor here, and this is your country. You've all been working with primates much longer than I have, and know more about how your country works than I do. I'll never forget that, and I want you to never forget that I'm here to support you, and help you become the best keepers you can be.'

Everyone seemed mildly reassured, which was a start, but we all knew actions would speak louder than words. I meant what I'd said, and I was proud to be part of what became a great team. As we all got comfortable with each other, the sanctuary started to run as smoothly as it had ever done. The staff were happy, and that made me happy.

Cercopan headquarters was a regular detached house off a quiet lane with a spacious wrap-around garden, albeit one filled with cages, surrounded by a high, crumbling concrete wall. Two large water tanks towered above the metal 'Welcome to Cercopan' gates at the front, and a short driveway led to a rattan-covered area that served as a sort of reception and education centre. Further down the slope, past more cages, you'd find the front door to 4 Ishie Lane, my home for the year.

During my entire stay we only had sporadic electricity. Power would often be gone for days (and on the odd occasion, weeks) on end, and there was no mains water. The telephone line only worked intermittently, and even when it did it was frequently disconnected by NITEL as a way of 'incentivising' us to pay our bill. Even then we had to bribe local officials to take our money, after which the line would be reconnected and we'd wait another couple of weeks before the whole sorry episode would start all over again. It was exhausting getting even the simplest things done, as Bob had said it would be. It didn't help that everywhere you looked it was clear that most of the city infrastructure hadn't seen any maintenance or improvement work in years, if ever. Given deficits in the public purse and the high levels of corruption, it was unlikely to get any better in the foreseeable future, either.

Water was a particular problem, and my biggest worry. Every week I would go, cap in hand, to one of a number of the larger construction companies in the city to beg for a water tanker to come round and fill our two large plastic tanks. Only once did I fail. On that occasion we had to collect water ourselves from a local hand pump, using dozens of jerry cans. Keeping the water flowing was by far one of the most stressful things I had to deal with, and I had to make friends with officials I'd not have given the time of day to in different circumstances. But I did okay. Our two big water tanks, once full, fed into the main house and to taps outside for the keepers to use so, although it was often a little cloudy and needed purifying, we did at least have water.

Mosquitoes were plentiful, attracted by the promise of captive primate blood and pools of water left over after enclosures were cleaned. It was only a matter of time before I got malaria, although I did hold out for a few months. This was something of a surprise given there was rarely a single moment, day or night, when I wasn't at risk of being heavily bitten. There's nothing more uncomfortable than having to continually cover your body with sticky insect repellent when you're already sweating profusely in the tropical heat. The primates came off particularly badly. Used to living out of reach higher up in the trees, when bitten some lost the ability to move their legs before slowly, lethargically having the life drained out of them. We didn't lose many that way and, although we suspected malaria, we never found out for sure. Regardless, such a slow death was particularly distressing to see. The condition became known as 'shuffle'.

I had my own room in the house, although I had to wait a couple of weeks for Suzanne to leave before I could lay claim to it. It was right at the back, and was hot, stuffy and dark, with little ventilation. You'd not dare open the windows, given what might fly in to join the cockroaches already camped out in the deepest, darkest corners. The furniture was old but solid, including the mattress which was a little firmer than I was used to. It reminded me of the kind of furniture display you'd find in one of those large, old, traditional, family-run department stores that used to sit on the high street years ago. But at least I had a room and, as I'd often remind myself, this wasn't a holiday. Life in the house was basic, with Eme the housekeeper doing an expert job of making sure everything ran smoothly. She fussed constantly over us, making sure we'd eaten enough and that we ate healthily. Eme was a gently spoken, kind woman, deeply religious and someone hard not to take an immediate liking to. Pounded yam and groundnut stew became a favourite of mine and, although Eme made it whenever I asked, I could see how much effort went into breaking up the yams. I tried not to ask for it too often.

Most of the other white people I met in Calabar worked in the oil or oil-related industries and lived in relative comfort. Over time, many of the Nigerian friends I made ended up feeling sorry for me living the way I did, even though it was my choice. Few could understand why I'd choose to live in a small room in a house on Ishie Lane, with a mosquito-infested garden full of monkeys, no running water and sporadic power, all while not earning a penny. It was sometimes hard to explain the importance of the conservation work being done at Cercopan and, more broadly, in Nigeria, but many appreciated our efforts and our willingness to help. Sometimes we (I and some of the other volunteers – Bob from America, and fellow Brit, Chris) would be invited to the local Chinese restaurant at the upmarket Mirage Hotel complex where we'd be offered free spring rolls and sweet and sour sauce with our beer. We had many good nights with our Nigerian friends. I wonder where many of them are today.

Calabar, the capital of Cross River State, shares a southern border with Cameroon and is a quiet, rather pleasant city that feels more like a large town. It had a population of around 350,000 back then, with Igbos and Efiks among the dominant ethnic groups. I was able to walk safely in the streets, even at night. In all my time I never witnessed a single fight or trouble of any kind. Life was hard for most people, but they went about it peacefully and with dignity. Given everything else going on around the country, it was like an island of calm in a stormy sea, and I felt lucky to be there when I thought of all the other places I could have ended up.

Thanks to the efforts of its enigmatic governor, Donald Duke, high-quality hotels had begun to spring up, along with shops that sold an increasing number of Western-style aspirational goods, albeit at a price. Ice cream had 'arrived' a couple of years earlier, and there were Chinese and Lebanese restaurants to choose from. Shwarmas were a particular favourite of ours, served from a hatch at the Mirage complex for 250 Naira (which was around $1.50 back then). It was only later, once we realised how expensive that was for a single meal, that we took our custom to more local places where a plate of omelette and spaghetti, or beans and rice, cost just N25. Food and food safety concerns aside, the conversations with the locals that accompanied those meals were priceless.

One morning, a few weeks after my arrival, Debo knocked on my door. Debo was the sanctuary vet, well-built and always well-dressed in a perfectly ironed shirt and trousers, whatever the weather. He was softly spoken, never raised his voice, and always seemed to be carrying work folders or a small leather satchel around with him. He'd not had the best of experiences with some of the earlier volunteers, and had decided to lay low while he tried to figure me out. He spent a lot of time reading quietly, studying and working alone in his lab. It was clear he was a talented individual, and was incredibly committed to his work. He was also well-liked and respected by the other keepers and sanctuary staff. Like many of the Nigerians I got to know, he was also deeply religious, and I enjoyed his calm and thoughtful approach to everything he did. There were occasions when I drew heavily on him, especially during times of stress, crisis or emergency. Debo was my Mr Dependable. I would later fund his relocation to the UK, where he would become a successful and talented – and fully qualified – vet.

I opened the door.

'Ken, we've had reports of a guenon being held near here, and we're arranging for it to be collected. Is the vehicle available, or do you think we could do this on a bike? And should I go alone, or should we go together?'

'What kind of guenon is it, Debo?'

'It's a juvenile mona,' he replied, 'so this shouldn't be too difficult, to be honest.'

And so began my first primate rescue in Calabar, although it turned out to be more of a confiscation, really, with the 'owner' willingly giving up the juvenile mona guenon he'd somehow come to possess. It wasn't entirely clear – it rarely was – but it was likely he'd killed the mother for food and then taken the infant in the hope of selling it as a pet. That was before someone had decided to report him. Given this was a rather straightforward operation, and juvenile monas were relatively small and manageable, I decided to go with Debo to collect it. Many of the more complex rescues, those involving greater primates such as chimpanzees, for example, or those with 'owners' who also happened to be members of the political or military elite, required a patient, sensitive, ego-massaging, non-combative approach.

So that morning, Debo and I headed off towards Oban Hills to collect the guenon, whom we'd later name Oban for obvious reasons. He had a broken arm, an injury likely sustained as he was pulled away from his dead mother, but he was otherwise in surprisingly good shape. We found him tied to a hot, corrugated metal roof with a short piece of string around his waist. Oban would sleep in a travel box in my room for the first few nights, with a soft toy as a substitute mother, while we slowly brought him back to health. Oban was a poor sleeper and I was woken many times each night by his screaming. I think he was having nightmares, which was no surprise given what he'd been through. I did my best to comfort him, but it never felt as though it was enough.

A few weeks later, after he'd settled down and put on a little weight, and we were happy that his arm wasn't going to give him trouble, he was introduced to the other mona guenons at the sanctuary, firstly in a smaller satellite cage and then into the main enclosure. It always felt like a small victory to reintroduce a rehabilitated primate to other primates, even if we knew the chances of them ever running wild again were slim. Captive life could never be a substitute for the wild existence many of them were used to, however well we looked after them.

Another rescue, a few months later, turned out to be far more challenging. I came across a red-eared guenon completely by chance at an entertainment complex a few of us had decided to visit one evening. We wandered between the tables in search of somewhere to sit, zigzagging in and out of the crowd to avoid the huge loudspeakers and the bright strobe lighting that pulsed around the large, open grounds. We found a space next to an old-fashioned, circular birdcage, and it was then that I spotted the guenon. The constant noise, flashes of light and comings and goings of people would have been mental torture for this young monkey. Pieces of kebab meat were lying on the cage floor, probably thrown in by one of the party-goers for a joke, and every few minutes someone would wander past and poke or bash the cage.

Checking that none of the bouncers was watching – we'd already been told not to touch anything – I slowly pushed my forefinger through the bars and the monkey's tiny hands gently clamped tightly around it, its cheeks pushing up alongside. As it held on, its eyes slowly closed as if, for the first time, it felt safe. Incredibly, despite all the noise, it fell asleep, more a sign of how exhausted it was than anything. We knew we'd have to rescue it, and soon. We discussed our limited options as we finished our drinks. Luckily, my Mr Dependable was there with us.

'So, Debo, what do you think?' I asked him.

'To be honest, if we don't get the guenon out, and soon, it's going to die. It's distressed and severely dehydrated', he replied. 'I think we need to get it some proper food and milk while we figure out how best to approach the owners here.'

My last act, that first evening, was to pull my finger gently away from the cage. The guenon gave out a piercing, distressing scream, frantically grabbing at me to put my finger back. It felt totally and utterly heartbreaking not to be able to do so.

For the next couple of evenings we returned, smuggling in sliced mango and specially-prepared milk in a tiny feeding bottle, doing our best not to draw the attention of the staff. In parallel with our efforts to simply keep the red-eared guenon alive, we'd started speaking to the owner of the entertainment complex, who happened to be related to Donald Duke, the Governor. It may have been illegal to keep primates as pets, but the entertainment complex had been set up as a kind of zoo with a variety of other animals on display, some of which, quite remarkably, turned out to be dead on closer inspection.

It took several phone calls and two in-person meetings to secure the release of the guenon. Discussions with the complex owners were friendly and amicable. After all, they held all the cards and had relatives in high places. Zena was in town and, given her role at the sanctuary, she led the negotiations. I attended but kept quiet. If Zena couldn't get the primate released, there was going to be little I could do. Debo was once again with us.

'So, we were here the other evening and noticed you had a monkey in a cage near the back of the complex', Zena began. 'We were sitting quite close and saw that it's very ill and in need of medical attention. If it doesn't get help it will almost certainly die.'

The complex owners didn't look too convinced. They knew we were from the sanctuary and it was our job to confiscate captive primates. To them, this sounded like an excuse to take ownership of it. Zena turned to Debo.

'Debo, how do you feel about the guenon?' she asked.

'Well, it looks severely dehydrated and is displaying signs of extreme stress', he began. 'And it's not getting the right sort of food, either, so I'm very concerned about its chances of surviving much longer out there.'

The owners said they'd have a think and get back to us. We didn't give them the chance, returning the next day. After a final conversation, where we stressed once again that the primate was close to death, they agreed to let us take it on one condition – that we return it as soon as it had been nursed back to health. Of course, once it was out of sight the owners soon lost interest and, fortunately for us and the guenon, in no time at all they had forgotten all about it.

Agbani, as the infant came to be known, was named after Agbani Darego, the Nigerian model who had just been crowned Miss World (a year later there were extensive religious-based riots in the north of the country, in protest of the holding of the 2002 competition in Abuja). After an extended period of treatment, and countless hours sleeping on my chest in those critical early days, she eventually recovered and was reintroduced to some of our other red-eared guenons, initially in a satellite cage and then, after everyone had got to know each other, into the main enclosure. Agbani settled in well but never fully recovered from her earlier torment. Sadly she only lived another year or so, passing away a few months after I'd left the country. Rescues rarely have truly happy endings.

By the time I'd hit the halfway stage of my Nigerian adventure I was in need of a break. Bob, our long-haired, laid-back American volunteer, had become a good friend and we discussed heading to Cameroon for a week. Bob came across as your typical American. Nothing seemed to phase him, and he enjoyed a cold beer whenever he could get one, as well as the odd cigarette. He never seemed to go anywhere without his red San Francisco 49ers baseball cap, and he would sit up late some nights playing Nirvana on the sanctuary cassette player. He was fun and good to be around, which was lucky given we only had each other's company most of the time.

Seeking adventure in its purest form, Bob and I opted to travel to Cameroon by sea. This was to be far from straightforward, and fraught with danger. Looking back, I'm amazed that we even attempted it. Scheduled services hadn't run for years, so the only option was to show up and look for a boat that was close to fully loaded, track down the captain and negotiate a ride. The first rule, we were told, was to only consider boats with two outboard motors in the likely event that one would fail while at sea (as ours did, funnily enough). 'Flying boats', as they are known, are long and thin, and usually piled high with cargo, so foot passengers are rarely accommodated. It's a method of travel not for the faint-hearted. It involves heading out into the Gulf of Guinea, into open seas, without proper life vests, and on an overloaded and likely poorly maintained boat, with a captain probably poorly qualified to make the journey. A crew of two would join Bob and me and, after a lengthy wait of a few hours, we finally headed out. Our first engine failed before we'd even left the mouth of the Calabar River. Whether the second would last the rest of the 100-odd mile journey to Douala was anybody's guess.

By some miracle it did, and after a short taxi ride our holiday could begin. For the next few days Bob and I had an unforgettable time travelling around the Cameroonian coast, swimming in the Atlantic Ocean while ducking the huge, lukewarm waves, with black sandy beaches, sweeping coastal forests and the looming shadow of Mount Cameroon as our backdrop. We lazed on sun loungers by hotel pools that we paid to get access to, and we drank cold beer and thought about how easy things were compared to Calabar just a little way up the coast. As a gentle reminder of what we'd left behind, we made the short journey by car to visit friends at the primate sanctuary in Limbe, where we shared experiences of primate conservation and life in the two countries. I was a little envious of how much easier things felt here, and how being close to the sea made everything seem so much calmer.

Our time passed surprisingly quickly and, before we knew it, we were heading back to Douala to attempt our earlier journey in reverse. We followed the same rules, looking for a twin-engined flying boat that was close to being fully loaded, and then joined a couple of other, well-dressed passengers, government officials we were told, for the journey back. If we thought we'd seen it all on the outward journey, nothing quite prepared us for what was to come on the way home.

As soon as we hit open seas the captain promptly cut the engines and threatened not to restart them until we paid an increased fare. Bob and I decided to keep a low profile given that the targets seemed to be the other Nigerians on the boat. After much shouting and arguing, and the eventual exchanging of money, we continued on our way. Not long after that, things took a turn for the worse when our crew spotted a boat speeding towards us from the coast.

'Oh, no! Here we go again', muttered Bob.

'What's the matter?' I asked.

'Look', he said, pointing off into the distance.

I followed his gaze, catching a speed boat out of the corner of my eye, heading in our direction. 'Who do you think they are?'

'No idea, Ken. But best we just lie low again, I think. That seemed to work last time.'

I hate to admit it, but I was more than a little anxious. We were, after all, pretty vulnerable, bobbing around in an overloaded boat, out in the open sea, still quite a distance from land and the relative safety of Calabar. Our captain cut our engines yet again and, as the chasing boat got ever closer, our fears were confirmed.

Bob leaned towards me. 'Guns, Ken. They've got guns.' We stayed at our end of the boat, staring out to sea.

'I don't think there's much we can do, other than stick to our plan, Bob', I whispered. So that's precisely what we did.

It's an odd feeling being so utterly at the mercy of people you don't know, and in a situation which could go any number of ways, none of which you control. I was also, perhaps strangely, acutely aware of Britain's colonial legacy each time I travelled to Africa. Most people liked the English, but not everyone gave us a warm welcome. Bob, being American, was in a worse position. Most people seemed to dislike America for reasons that weren't always clear to me. My time in Uganda, in 1998, had coincided with bomb attacks on the US Embassies in Dar es Salaam and Nairobi, and the Americans in our group were understandably nervous.

Bob and I were nervous right then, too. We were quite far out, and the only people within miles were a bunch of masked strangers who were about to pull up in their boat. They could choose to shoot us there and then, and nobody would know. We'd become another headline, or one of those unexplained murder mysteries, destined for cable television, never to be solved.

Their engine idled and the men's boat drifted the last few metres towards us, easing up alongside. Suddenly everything was quiet, except for the sound of the waves lapping up against our boat and the flapping of the plastic cargo covers in the wind. Within seconds the shouting started. Once again, the targets appeared to be the other Nigerian passengers and, this time, the crew. Things were tense for what felt like an age. More money was exchanged. Words were exchanged. Guns were waved around. And then, before we knew it, the men throttled up their engine and were heading back towards the coast. We were very quickly on our way again, too.

'That went well', I said to Bob, with a mixture of relief and gratitude. We had no idea what had just happened, but we seemed to have survived it.

Within an hour we finally reached Calabar. We didn't arrive back in the harbour as we were expecting, though, but instead dropped anchor next to a grassy bank, somewhere on the outskirts of the city, where we were met by what appeared to be friends of the crew. One of them looked like he may have been a customs official, and had a pistol. Our fellow passengers climbed up the bank and scuttled off, and we were encouraged to follow.

'Your journey ends here, my friends', shouted the captain. Bob and I looked at each other, blankly.
'Here? Where are we?' I asked.
'Calabar!' he replied. 'We can't get into the harbour so we're dropping you here. Off you get.' Not for the first time, Bob and I decided it was best to just do as we were told.

So off we climbed, grateful that we'd made it back to dry land at all. It wouldn't be until a few weeks later that our decision would get us in trouble, immigration officials asking us to explain why our passports had not been officially stamped when we returned to the country. As with the many other challenges we faced during my time in Nigeria, we eventually managed to worm our way out of this predicament, just like all the others. Nothing was ever, ever straightforward.

If I'm honest, most of my time in the country was spent in survival mode. I'd navigated most things relatively unscathed, but my luck was about to change. Shortly after my 36th birthday I invited all of the sanctuary staff out to a local bar off Odukpani Road to celebrate. Odukpani Road is a major highway running south to north through the entire length of Calabar and beyond, and was only a 15-minute drive from Ishie Lane.

We arrived at a bar to find a small crowd in party mood gathered outside. The scene was lively, with music blaring in all directions from competing sound systems, disco lights flashing, street food sizzling on barbecues, and people darting in and out of a constant stream of motorbikes amid the smoke and flickering lights. Cameroon it was not, but I always enjoyed spending time away from the sanctuary with friends and colleagues, and the fact they were all with me that night was all that mattered.

In fact, I enjoyed my time socialising with the staff, despite being advised not to. 'They'll only take advantage of you if you get close to them' was the usual warning from the European expats I met. It was a warning I generally ignored. That night, as we sat drinking our beer and enjoying the street food, the staff gave me several tins of tuna as a belated birthday present. They were always concerned about me getting ill and needing to be strong, and they knew I liked tinned tuna. It was something of a delicacy compared to the cheaper local fish they preferred. This was their way of showing how much they cared and, to be honest, tuna probably was the best thing they could have got me at that stage of my trip given the amount of weight I had lost.

At the end of the evening Jerry, the education programme co-ordinator, offered me and Chris, a fellow Brit who had replaced Bob a few weeks earlier, a ride back to Ishie Lane on his motorbike. Three people on a bike really was nothing and, if anything, it was a fairly light load. We'd all ridden together like this many times before, sometimes with helmets, when available, but mostly without. That night was nothing unusual. We weaved our way between what remained of the people and traffic on the side street, and headed back towards Odukpani Road.

Full details remain sketchy – your mind sometimes does this when you experience this kind of trauma – but as we picked up speed I remember turning to Jerry.

'Thanks for offering to drive us both back, Jerry.' As I spoke, he turned his head slightly so that he could hear me.

'It's the least I could do, Ken. You'd both struggle to get a bike this time of night.'

Looking forward, past the side of his face as he spoke, I had a full view of the road ahead. I remember seeing a faint set of car lights creep out from a side road, wondering if it was going to stop. Thinking about it now, I was strangely calm given what was clearly about to happen. By the time Jerry turned his attention back to the road ahead it was too late. The car hadn't stopped, and we slammed straight into its front left wing, catapulting us through the air. The last thing I remember was the sound of the bike engine revving up as our back wheel left the ground. What happened to the bike after that remains a mystery.

After the sound of the engine revving, there was nothing. No sound, no vision. Just darkness. It's funny what the mind can do in situations like this, switching off entirely or cleverly erasing memories of traumatic events, perhaps to save us from reliving them over and over again. For that, I am grateful, even if I do sometimes wish I could remember a little more about what happened to me in the seconds following our collision.

Chris, who was seated behind me, ended up furthest down the road. Then there was me, and then Jerry. Despite travelling the least distance, Jerry was most severely hurt. Chris was shaken but, remarkably, came out largely unscathed. I remember coming round after what felt like a few seconds, lying on my back and looking up at the dark sky, the warmth of the road giving me little comfort. Tins of tuna were scattered around me. Given we were all now blocking the road, traffic quickly built up behind us, and I could hear the commotion as drivers debated how they might help us, or the best way to drive around us. Others were trying to gather up the tuna, no doubt with the next day's dinner in mind. The Nigerian Police arrived surprisingly quickly, but their priority was clearly with other road users as they attempted to drag us out of the way. By that time I already knew my leg was broken after I'd tried to stand, only to see it bend awkwardly sideways. Although the pain was yet to fully kick in, I decided to scream, hoping the police would leave me alone if I did it loudly enough. I did. And they did.

There wasn't much to be grateful for, but I was happy to be conscious and able to maintain a level of control over the situation, and to have only my leg to worry about. Other minor cuts, bruises and grazes to my arms and body were little bother in comparison. It really could have been so much worse, particularly considering our lack of protective gear. My first thought was how to avoid being taken to the local hospital, where any treatment was likely to be hit and miss, and largely out of my hands. By a stroke of luck, the local area manager for MTN, the mobile network, was heading the other way and pulled over to see what the commotion was all about. In no time at all he stepped in and his silver Mercedes quickly became my ambulance. With a little help I was gently laid across the back seat, my leg delicately pushed up against an armrest, and I was on my way to a clinic. Jerry was in another car, whose I don't remember, following closely behind.

By the time we arrived the pain was seriously kicking in. I had heavy, beige, ankle-height walking boots on that night, and the clinic's efforts to take my right boot off while my foot flopped around was excruciating. A puncture wound, caused by one of my broken bones bursting through the skin, randomly spurted blood as my foot was moved. A shot of Ketamine, an anaesthetic largely reserved for veterinary use these days, killed the pain for a short while but sent me on wild, hallucinogenic trips known as k-holes, which felt like nightmares with no end. I imagined I was clinging to the outside of a rollercoaster, being flung around left, right and centre. Then the torment would stop, tricking me into thinking I'd woken up, and then start all over again. And again. I have no idea how long this nightmare went on – it felt like forever – but when I did eventually wake my shoe and bloody sock had been successfully removed. My lower leg was now secured in a locally made wooden splint, with bent nails hammered down between the joins and the corners. The state of the makeshift splint was the least of my concerns, although it did bring an unexpected and welcome smile to my face. I was simply grateful that it was doing its job, and my leg was now resting in a position and at an angle where the broken bones were least likely to grind together every time I made the slightest move.

For the next couple of days the management at Cercopan tried to figure out what to do next. A serious accident like this, involving an overseas volunteer, had never happened before. For obvious reasons the insurance company that would be footing the bill for my treatment wanted me back somewhere where they'd incur the least cost. In my case, that meant a return to Jersey, where it would be free. The first stage of my journey home took place three days after the accident, a private charter flight from Calabar to another clinic in Lagos, where I spent a further three days while my final flight was arranged. The clinic was clean, organised and comfortable, and the food was some of the best I'd had in months. Before I left, my wooden splint, still complete with its array of bent nails, was removed and replaced with a plaster cast with the help of about a dozen Nigerian medical students. I'd been asked if I was happy to have them there to help, and I didn't see any reason why not. By the looks on their faces, I don't think they had been given the chance to treat many white people before.

This wasn't the end of my Nigerian adventure that I was expecting, or hoped for, but at least I was now safe, well looked after, and comfortable. As for Jerry, he would remain in Calabar and eventually have his own legs fixed there. He would spend the next couple of years on crutches, slowly recovering and eventually returning to work.

My final farewell took place through the back of an ambulance window as we zigzagged in and out of the heavy, early evening traffic that clogged the main road to the airport. Everything looked and felt the same as it had on my arrival almost a year earlier. The streets were crowded, *matatus* picked up and dropped off passengers, street food sizzled away, smoke rose into the darkening skies. Ten hours later I'd be in London, back in the place where my journey had begun.

Another five hours after that and I was finally back home in Jersey, arriving at the airport to be greeted by an ambulance. Six years earlier I'd left the island, full of optimism, in search of meaning and purpose. A lot had happened in that time: a successful stint at university, a team leader's position at Cable & Wireless, summer jobs with a language school in Brighton, time spent volunteering in Zambia and Uganda, a year of teaching English in Finland and, of course, a year in Nigeria. Despite the variety, quantity and craziness of it all, I didn't return home with any answers. If anything, I'd gone backwards and not forwards, now finding myself unexpectedly homeless, with no job, no money and no idea what to do next. Not only was I back in the place I'd spent years trying to leave, I now had a broken leg to deal with, and a recovery that would take the best part of twelve months.

Although I wouldn't know it for a while, Nigeria wasn't done with me yet. In April 2007 a text messaging platform I was yet to develop would be used to monitor their presidential elections. It would make international news, turning my world, and my life's purpose, completely upside down in the process.

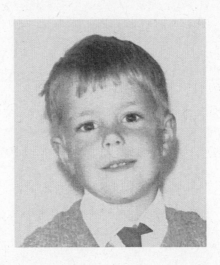

Above, a photo of me in my school uniform, probably taken when
I was around seven or eight years old. Below, number 73, the house on Five
Oaks Estate where I spent all of my childhood and part of my early adult life.

THE LEARNING CENTRE.

LA HOUGUETTE · FIVE OAKS ST, SAVIOUR · JERSEY · CHANNEL ISLANDS · TELEPHONE: 0534 30062-53935

To whom it may concern.   KENNETH BANKS.

Kenneth Banks has worked at the Learning Centre on
a CBM 4032 computer for about one year during which
he appears to have achieved a considerable mastery
of programming techniques in BASIC. He has carried out
advanced work on Computer Assisted Learning (CAL)
teaching programmes for the Centre. (Part of our work
involves the use of CAL techniques and research in to
these applied to academic learning difficulties. We are
undertaking some of the most specialized work in this
field in the U.K.,if not in Europe.)
His approach to his work is conscientious,enquiring and
intelligent and he shows distinct ability to persevere
in working out a computing problem. When necessary,he is
able to refer to and assimilate research literature with
subsequent practical application on the computers.
While he is still relatively young my view would be that
he could go far in having a successful and rewarding career
in computer programming.

November 1982.   F.R.V.Cooper MA,B.COMM.TCD.
                                  Director.

Above, my reference from Mr Cooper extolling the virtues of my programming
skills on his Commodore PET computer at Club. Below, the mainframe computer
I took charge of a few years later at Hambros Bank as my IT career took off.

MUM

In the very unlikely event that something happens here's a few things that I think you ought to know.

On the official side, I've got lots of policies with various companies - the two firms over here that I deal with are Actus (66620) and Homebuyers (58355). Have a great time spending the money (but spend yours first, eh?).

Thanks for being a great Mum through thick or thin, good or bad - I hope that my Mum in my next life is like you (except that I hope she gets her inheritance a little before you got yours!). If anything has happened, don't be sad - these things happen! We've all been brought up slightly out-of-the-ordinary (bacon and spaghetti for breakfast, etc.) and we all have ambitions, visions and take risks because you have brought us up well with a good, healthy attitude to life. I think that we've got an exceptional family and if it has come to a premature end then I've already had more than a lifetime's worth of love and happiness from you and the rest of the family.

I always like to have the last word, and I've done it again! Don't forget that I love you all very much, but don't be sad - I'll save you a space on my cloud and see you again.

Lots of love, always, to everyone.

The note I left for my mum on the coffee table of my flat in July 1993.
Not knowing how that very first trip to Zambia was going to go, I wanted to
cushion the blow if anything happened to me.

The train article, in the *Jersey Evening Post*, that drew my attention to
the Jersey Overseas Aid project in Chilubula, Zambia, that summer. That trip
would become my first taste of overseas aid work. (Reproduced with permission.)

Above, the painting of lions on wood that I bought from Justice
Kabango on that first Zambian trip. Below, Chiti, his son – the quiet boy
in the crowd – taken in the compound where we stayed, August 1993.

Above, Andrew and I heading back into the village after our Mount Elgon climb.
Taken at the midway point of my trip to Uganda, in 1998. Below, departing
Jersey Airport for Zambia in July 1993.

Above, the building site in Chilubula, Zambia, 1993. By the time we left, the window frames were in and the roof was about to go on. Below, with Oban, my first primate rescue. Taken in Calabar, Nigeria, January 2003.

What could so very nearly have been my last ever photo. Taken
as we began our whitewater rafting adventure on the Zambezi River.
Note the steep cliffs leading up from the water, and the rope along the top
of the boat, just behind me, that we were all meant to keep hold of.
Taken in August 1993.

Part II

Purpose Lived

Ring Ring

Once more, a kick in the teeth. That was my assessment of it, anyway, as I lay in bed that day in late 2002 with no obvious reason to get up. Void of ideas and options, with no home, no job, no money and little mobility, everything felt empty, the path ahead unclear. I had no idea what to do next, no idea what else I could try. Life had come full-circle and I was back in Jersey, back where it had all started. I felt I'd done everything I could to seek out the meaning and purpose that seemed so important in my life. I'd taken all sorts of risks, embraced financial insecurity, sold up and started over multiple times, run off to different countries and taken paths that felt right at the time, but had led me nowhere. For someone who'd always felt his destiny lay in his own hands, ending up here with so little reward was a depressing, lonely, difficult place to be.

That morning my bout of self pity was interrupted by the sound of the phone ringing in the hallway below. I dragged myself out of bed, made it gingerly down the stairs, and picked it up.

'Hi Ken, it's Karen.'

'Oh, hi there! What a lovely surprise. How are you?'

Towards the end of my time working at Jersey Zoo I'd met Karen Hayes. While I was running around between university in England, English teaching in Finland and primate rehabilitation in Nigeria, she'd been hard at work at university producing one of the earliest studies on the impact of coltan mining in the Democratic Republic of Congo. The insatiable demand for coltan, a mineral used in all manner of digital technologies, particularly mobile phones, was taking a terrible toll on the environment and the local populations who were mining it, often by hand.

'So, Ken, how are things?'

'Oh, well, not too bad', I replied, briefly sharing what I'd been up to since the last time we'd met, finishing with my accident in Nigeria. Karen was happy that I was now safe, on the mend and, as it turns out, available.

'Listen. I have some new work coming up that I think you might be interested in.' I was all ears.

'I'm now running the Corporate Team at Fauna & Flora International (FFI) in Cambridge, and we've been awarded a grant by the Vodafone Group Foundation.'

'Sounds great, Karen. Congratulations!'

'And this is where I think you might come in. I need someone who's into conservation, has experience working in Africa, and who understands technology. You're the only person I know with this kind of background.'

Karen and her small team at FFI were interested in how mobile phones might help conservation and development work across Africa, and how they might also be used closer to home to promote the work of organisations like theirs. The project was funded for eleven months with a generous £750,000 grant. Researching and publishing a report, the first of its kind, on top of designing, building and launching an innovative online mobile service, all in just under a year, was quite an ask.

That short phone call put paid to any illusion that I was ever in control of my own destiny. It wasn't lost on me then, and it hasn't been since, that I wouldn't have been in Jersey to take it if I'd not broken my leg in Nigeria a few weeks earlier. This would be one career break to beat them all. Seemingly out of nowhere, all of the random pieces of a puzzle I'd been haphazardly and painstakingly putting together since Live Aid in 1985 fell into place. Was this the stroke of luck that the universe owed me? Had it decided I'd suffered enough? Looking back now, I was pretty close to rock bottom. After a brief spell down in the depths, wallowing in self pity, I was now about to take my first tentative steps back up.

A couple of weeks after our call I hobbled my way to Jersey Airport on crutches, ready to take a flight to Southampton, the closest airport to where Karen lived. Karen's partner was Simon Hicks, who'd been Trust Secretary at Jersey Zoo for over twenty years, and it was thanks to him that I'd met Karen in the first place. Simon was a tall, friendly, welcoming, well-spoken character, the sort of person you wouldn't be surprised to hear had been in the army in his younger days. He also had a tremendous sense of humour and a booming laugh. Often dressed in casual, outdoor clothing, topped off with a flat, green cap that covered his light, wispy hair, his passion for conservation was infectious, and his knowledge of wildlife second-to-none. Simon and I had a tremendous working relationship, and I was grateful for his trust and support all those years earlier as I built out the membership system at the zoo, aptly named Jambo after the zoo's famous gorilla resident.

I arrived in the UK for what turned out to be a rather unconventional job interview. That night the three of us went out for dinner, discussed the role, caught up on the previous few years of our lives, and plotted a path ahead. It wasn't a given that I'd get the role, of course, although on paper I was the ideal candidate. I thought it was a dream job, one that combined everything I loved or was good at – technology, conservation, development and Africa, a continent that had captivated me and that I was desperate to return to.

Because of the relatively short timeframe for the work, the Project Technical Advisor would be hired as a consultant, not a regular member of staff, and the generous day rate in the budget would go a long way to making up for all the income I'd lost over the previous few years. I remember going to bed that night unable to sleep, thinking that I'd be earning nearly as much in a day as I'd earnt in a week at Cable & Wireless. I still needed to convince myself that I was able to take on such an important and ambitious piece of work, but at least the opportunity was there. Despite everything spinning around in my head, I managed to drift slowly into a deep, easy sleep.

The final hurdle was to meet Mark Rose, the CEO of FFI. As Karen put it, 'We need to make sure he likes you'. The next day Karen and I took the train to Cambridge where I'd meet Mark. It was a long journey, via London, and Karen worked the whole way. I'd soon learn that this was nothing unusual. I was still technically unemployed so I didn't have much to do, so I just stared out of the window and fretted about the conversation with Mark that would ultimately decide my fate.

Karen's work ethic was incredible. An instantly likeable, wavy red-haired Irish woman, she was always busy doing something. She was a also brilliant presenter and would turn out to be by far the greatest networker I'd ever meet, too, lighting up a room as she wandered through it, making everyone she approached feel immediately at ease. That was all to be a taste of things to come, of course, and I wasn't there yet. I felt so close to achieving the impossible, finally finding and living out some kind of purpose or meaning. There was just that short conversation with Mark in the way.

FFI were on the second floor of an uninspiring-looking shared office building next to Cambridge train station in what is now a Microsoft campus building. A few years after my work ended there they bought it and had the whole thing knocked down. Shortly after that, it re-emerged as something more modern and beautiful. Back in 2002, when I visited for the first time, the area was surprisingly messy, under-developed and run down. Between the train station and FFI's offices was a large patch of fenced-off wasteland which appeared to be the site of another building that had been knocked down and cleared years earlier, and the space just left to fill with weeds and litter. It's a very different picture today with cafes, shops and bars on almost every corner.

I remember that first visit to Cambridge well. FFI weren't just any old conservation organisation, but the oldest and one of the most respected in the world. They went about their work with little fanfare, and with little ego, choosing to work through local partners instead of bulldozing their way in and splashing their name and logo everywhere. Their focus on local communities was also refreshing and appealing and, morally and ethically, they were the perfect fit for me after all I'd seen and learnt over the previous few years.

I stepped into Mark's office which, at that time, was in a quiet corner of the building overlooking the front of the car park. Carvings and paintings dotted the shelves and walls, even the floor, giving a wonderful flavour not just of FFI's work but also of Mark himself. A zoologist by training, he had worked early on in his career in remote parts of Africa and Asia-Pacific. When I first met him he'd only been CEO for a few years but, even then, he was well on the way to transforming the organisation from one with just a handful of active projects into a global outfit with close to a hundred in over forty countries. Building out the corporate work, which Karen was now leading on, was very much part of his vision for the organisation back then, and the Vodafone work particularly excited him.

'Oh, hi Ken. Come in and grab a seat. I won't be a minute.' Mark disappeared for a few moments, giving me my first chance to study all the mementos of his travels, and FFI's wider work, scattered around the room. He quickly returned.

'I hear you had a nasty accident recently, and saw you limping as you came in. How's the recovery going?' I told him about my accident in Nigeria, and the work I'd been happily doing there just a few months earlier. Immediately we hit it off. I may have been no match for Mark and his own travels and experiences, but he clearly warmed to people who took risks doing things they loved, particularly in conservation. 'So, Karen tells me you'll be joining us to work on the Vodafone project, then?'

'I hope so, Mark. It's a really exciting piece of work, and it's a wonderful mix of technology and conservation, both things I've been involved in, and been passionate about, for a long time.'

We didn't chat for long. Mark always seemed to have dozens of things going on all at once but, suffice to say, the meeting went well and I really didn't have much to worry about. He clearly trusted Karen's judgement, and what was good for her was good for him. In no time at all I was back in Jersey for Christmas and new year, with a freshly signed contract which would see me back in Cambridge to start work by the middle of January 2003. A new year, and an exciting new beginning.

Despite all my self-doubt and seemingly endless conversations with my mum, I did get on the plane and I did arrive in Cambridge as planned. Yet another fresh start beckoned. Fortunately, by now I was used to moving around and living out of suitcases, and that's precisely what I'd be doing for the first couple of weeks while I found a place to live. For now, a single room in a simple guesthouse off Tenison Road, a short walk from the FFI offices, would have to do. During those first few nervy days, cooked breakfasts made up for the inconvenience, loneliness and new-job nerves. By the end of my first week I'd just about settled into my temporary home, got a good feel for my surroundings, become familiar with the FFI office and met the two colleagues I'd be working with for the rest of the year.

Of the three of us, Bill was the only one without any experience in the developing world. A tall, friendly, well-spoken character with short, curly brown hair and a fondness for woolly jumpers, Bill's background was more in local and national conservation issues, and he would support the project administratively. Richard was a shorter, bespectacled and bubbly (if not sometimes overly serious) character, a seasoned consultant and by far the most experienced of all of us. In fact, I'd go as far as saying he was the first professional global development expert I ever worked with. Richard was a sensitive, thoughtful guy who took his time before diving in with ideas or opinions. You always knew where you stood with him, and I liked that. He also taught me a huge amount about consulting, without ever realising it, given his organisational skills, work ethic and general attitude to anything he took on.

Richard and I were to get on well over the year and become friends. He had little to no technical experience, which he openly admitted, and I had little to no primary research experience (at this level, anyway), so we complemented each other perfectly. Karen knew that, of course, and she did a brilliant job carefully assembling a team that would end up over-delivering by the end of the year.

The work we were about to undertake at FFI coincided with the start of an exciting new era in what was, back then, the fledgling world of mobile phones. By the time I arrived in Cambridge a new breed of devices were only just beginning to hit the shelves, ones that could handle data (albeit at a crawl), that had colour screens, the ability to play fancy ringtones and games, and that could connect to the Internet (of sorts). Four months earlier, with much fanfare, Vodafone had launched Vodafone live!, a new multimedia service that allowed people to access pre-approved content through a range of pre-approved phones. Content providers, of which we would become one by the end of the year, would need to meet a strict range of technical requirements in order to serve up content to Vodafone live! users. It was only possible to become a content provider if Vodafone let you, but the nature of this closed system meant that, if you had decent content and could get visibility for it, there was serious money to be made. We would later find that out for ourselves.

I'd never owned a phone with anything like the capabilities of a Vodafone live! handset before. While I was still in Jersey I'd read up on the service and the phones that were supported – only two at launch, incidentally – so although I knew a little it was no way enough to be able to build anything that might work on one. We needed to get up close and personal with the service, so one afternoon Bill, Richard and I paid a visit to one of the local Vodafone stores, leaving a short while later as proud owners of the latest Sharp handset, the GX10. Looking back it was terribly crude, clunky, unfriendly and ridiculously slow, but at the same time amazing. It would be another four years before the iPhone would launch, changing the mobile landscape forever.

We quickly got up to speed with what Karen had promised in the project proposal, and started laying the foundations for the report we'd have to research and write, and the Vodafone live! service we'd have to build and launch. At the time it was mostly larger companies jumping on the mobile bandwagon. Those with technical knowhow, buckets of money and content that people would be willing to pay for were at a clear advantage. Unsurprisingly FFI, a one-hundred year old conservation organisation, wasn't one of them. For starters we didn't have much by way of technical resources – that was my role to fill, and I'd never built a mobile service before – and although we had some money it was a modest amount given our ambitions. We had very little content either, other than an incredible picture archive taken by FFI's talented photographer, Juan Pablo Moreiras. But it wasn't all bad, though. We had a great relationship with the Vodafone Group Foundation, not to mention first-mover advantage in the charity sector, and the sort of conservation stories people couldn't get enough of. We just had to fill in a few gaps, and find a way to string it all together.

The Vodafone Group Foundation was based in the main Vodafone offices nestled in the rolling hills of Newbury, Berkshire. It was an hour or so out of London by train, and Karen, Bill, Richard and I would make regular trips to meet up with our funder to request help, share progress, or generally build relationships. Meetings were always incredibly relaxed and informal, something all of us appreciated given what we'd taken on. Before each trip the four of us would rendezvous in the often freezing, cavernous concourse of Paddington Station, grabbing hot coffee and croissants before heading out on our train, planning and plotting as we went.

Newbury wasn't the only place on our travel list. We had a report to write, and South Africa and Mozambique became familiar stops as we tried to make sense of, and document, the slow but emerging mobile 'revolution' across the African continent. Back in 2003, other than a few stories here and there, little was known about the impact that mobiles were having in the developing world, and there had been no studies of any real significance that focused on the potential of mobile phones for conservation and development. When we arrived on that first trip, Africa boasted only six phones per one hundred people compared to over fifty at that time in Europe.

Network coverage was patchy and mobiles were expensive, giving people few reasons to own one. Evidence of how useful they were was largely anecdotal, with stories of farmers using phones to check prices at different local markets proving particularly popular. As we'd find out for ourselves during our early visits, phones were indeed having a very real impact on people's lives, even in those early days, and I was incredibly fortunate to be around to witness and document a small part of it.

Back in Cambridge we quickly developed a plan of attack for the new Vodafone live! service. Our first challenge was how and where to find high quality conservation-themed content. We also needed a technical partner who could build the site, and host the content, to Vodafone's very specific requirements. Not everyone could do that back then, and certainly not within our budget. And, if that wasn't enough for our small team, we also needed a regular website to promote the service to the general public. Although mobile subscribers were our primary audience, and Vodafone ones in particular, we were mindful to exclude as few people as possible. As a charity, FFI needed to leverage this opportunity to raise as much interest and awareness about its work as it could. Project management of the content sourcing and creation, and the building of both the Vodafone live! and the main sites, would be down to me. Both were tasks I would relish.

FFI's photographer popped over to Cambridge from Spain in the spring, where he spent what little time he had between trips to sit down with us. A modest, friendly, larger-than-life, often humorous man with a thick Spanish accent, Juan took the most incredible photos in ways that only truly talented photographers can. He saw angles, facial expressions, compositions, depths of field and juxtapositions that would pass most of us by. The subject matter of his photos certainly helped ramp up their appeal, of course, and he'd regularly be pointing his lens at everything from stunning wildlife to expansive landscapes. To say I was envious was an understatement, but it did mean we had some of the best wildlife photography around to turn into wallpaper images for the phones.

In the years long before the kind of high-resolution cameras we're used to seeing today, people would often buy images to save as backgrounds on their phones. In 2003, mobiles had not long boasted colour displays, and people were desperate to customise them as much as possible. Wallpaper images became a highly lucrative market, and we were perfectly positioned to sell wildlife images to support the kind of conservation efforts that were the subject of Juan's photos in the first place.

Ringtones were another huge opportunity. The Sharp GX10 that we'd all proudly walked out of the Vodafone store with was one of the first to support what were known as polyphonic sounds, meaning they could play two or more notes at the same time. Earlier phones made one sound at a time, usually a series of single beeps to signal an incoming text message, or a monotone sound for an incoming call. Attempts to make anything sound like an actual tune usually fell well short, and most were totally unrecognisable, and certainly not cool or worth spending any money on.

Although it was possible to select different tones for different functions from the phone menu, the choices were less than inspiring and most people stuck with the default sounds. For consumers desperate to make their phones stand out, though, polyphonic ringtones let you make use of almost any sound or tune with a richness that had previously been impossible. Unsurprisingly, record companies couldn't get enough of this new opportunity, and the vast majority of mobile owners willingly parted with their money to have their phone play a favourite pop tune when it rang. We knew this was an opportunity we'd need to grab, too.

As well as music there was a parallel market for 'alternative' sounds. A year after our own work began, a ringtone known as Crazy Frog became the first piece of mobile content to go truly viral, pulling in over 50 million downloads around the world. Released in 2004 by mobile content provider Jamster, sounds of a young Daniel Malmedahl attempting to make motorcycle engine noises made the company a reported half a billion dollars, and inspired music CDs, a game for the Playstation console and a range of merchandise including books and t-shirts. You can still find Crazy Frog on today's most popular music streaming services, and it sounds as awful as ever. While our ambitions were far more modest, alternative ringtones were a market ripe for exploitation, assuming we could get hold of conservation-themed sounds to do something with.

The British Library, of all people, came to our rescue. Housed previously in the British Institute of Recorded Sound, they had one of the largest collections of sounds in the world, including all manner of animal calls. I can't remember precisely how or when we got in touch with them, or even who told us about the archive in the first place, but several meetings and conversations later we'd licensed around a dozen spectacular animal sounds to use as ringtones. We still had plenty of work to do to get them to play on a phone, but we felt we were onto a winner with the distinctive zebra, gibbon, gorilla, penguin and tiger calls, among others. Nobody else was doing this back then.

By now we'd completed the almost impossible task of narrowing Juan's extensive photo collection down to about a dozen images – a mixture of mostly animal close-ups, and mountain and floral landscapes – and we were now experimenting with cropping and resizing these to fit on the growing range of Vodafone live! handsets. What worked for one didn't always work for another, and we were heavily restricted by the size of the image files. Network data speeds were considered quite fast back then, but these were pre-3G days and, looking back, it's hard to believe it was possible to transmit anything over the air to a waiting phone. We spent a lot of time getting these wallpaper images just right, and Juan made sure we presented his work in the best possible light. Nothing would get past him, and we were grateful for his attention to detail, as were the tens of thousands of people who would later end up proudly displaying his images on their phones.

By this time we were beginning to feel pretty pleased with ourselves, but there was still one other piece of crucial content we were missing. Games. Thanks to something called Java, a platform which happened to run on all Vodafone live! handsets, mobile gaming was beginning to take off, and we believed we had an opportunity to build something exciting, innovative and new. But before we could do any of that we'd need to come up with an idea for a conservation game and then, if it turned out to be remotely viable, figure out how to build it. Unlike the relatively simple task of picking a few photos or animal sounds from a library, games would be a different proposition altogether. Few charities, if any, were conceiving, designing and building mobile games back then. Few companies were, either.

As it happens, Lady Luck struck once more. One of Simon's daughters had been dating a guy called Tom, who happened to be co-founder of a young startup called Masabi. Today, Masabi is a hugely successful company focusing on electronic ticketing for the travel industry, but back then they were dabbling in gaming and random mobile apps, and working out of a mixture of bedrooms, lounges and borrowed office space.

They were always on the lookout for interesting things to do, always needed a little extra money, and were perfectly qualified to work on a game with us. We grabbed the opportunity to work with them, and Tom and his co-founder Ben didn't let us down.

We decided this might be a great opportunity to involve some of the FFI staff in what we were doing. Many were excited by the 'strange little mobile project being run by the corporate team' and, while many had volunteered to help, we were yet to take any of them up on their offers. Crowdsourcing (or perhaps that should be 'staffsourcing') ideas that could be used for a conservation game seemed like a great use of their passion, talent and energy, not to mention a lot of fun, and likely the best chance we'd have of coming up with something remotely useful, plausible or interesting.

Late one Wednesday afternoon we invited Tom to Cambridge. We shipped in a few bottles of wine and some beer, and a few bags of peanuts and crisps and gathered anyone and everyone with nothing better to do into our small office. We gave a rundown of the project which, at Simon's suggestion, was now called wildlive!, and Tom spoke a little about the gaming opportunity. And then we just, well, threw it all open. And as the wine flowed, so did the ideas. Most were crazy, or interesting but totally unworkable.

'So, how about this?' chimed in a colleague from the finance team. 'You're an animal, and you've lost all your body parts, and you have to move around a maze or world and find them all.'

Thankfully, after the ripples of laughter had died down, that particular idea was quickly kicked into touch, as were many others which were not, if we're honest, the kind of thing we were looking for. A couple of hours later, though, we did end up with the seed of an idea, and Tom gave it the green light from a technical point of view. Incredibly, a random gathering of conservationists, technologists, accountants and consultants had managed to come up with an idea for what would be the first conservation-themed mobile game. Silverback, as it became known, would become a hit not just among conservation fans, but also within the wider mobile gaming community.

Thanks to a healthy mixture of good luck and good planning, things were beginning to fall into place. We had the wallpaper images and ringtones in production, and coding for the game was very quickly underway. We'd found another company called

In-Tuition who were building the wildlive! mobile site (this is the one which would appear on the Vodafone live! screens) and a separate, regular website for public consumption was being written by another company. Along with the downloadable content we also started putting together website sections for latest conservation news, a members' area, and live field diaries, the first of which would come from an FFI tiger conservation project in Indonesia. Despite a standing start in mid-January, by the summer we were well on track to deliver one significant part of the project.

Of course, we still had a paper to research and write, not an insignificant undertaking in itself. While I'd been focusing largely on the mobile work, Richard had been busy with the report, and he'd also made great progress. We now had an outline of what we wanted to cover, and began to sift through the growing number of stories describing all the wonderful things mobile technology could do. That wouldn't be enough on its own, though, and we needed to plan a couple of trips to South Africa ourselves to get a better sense of how accurate these stories really were, and what this growth in mobile phone usage actually looked like. At the same time, we were keen to explore ways that mobile technology might benefit some of the project activities of FFI's main African partner organisation, Resource Africa, with a view to turning these into a project in their own right a little later.

Resource Africa's roots lay in early efforts to incorporate local communities into conservation initiatives, something that first began in the 1980s. Throughout the 1990s they championed an approach centred around the sustainable use of natural resources, with communities at the heart of their efforts. Resource Africa's approach closely echoed that of FFI, and the two had become close partners.

Our itineraries were dominated by field visits to community farming projects, women's economic empowerment initiatives and potential ecotourism sites. In South Africa, most of these were based in and around Bushbuckridge, an area straddling the southern border of Kruger National Park and the place where Resource Africa had decided, early on, to focus most of their work. We also made trips to Mozambique, in particular Maputo Elephant Reserve, one of the most beautiful places I've ever had the pleasure to work. The drive from Johannesburg to Maputo took the best part of a day but was well worth the effort, taking us through some quite spectacular scenery along the way. Going overland beats flying, any day.

During our visits we looked at how mobile phones might support or scale some of the conservation work being done there. A number of communities, for example, were producing souvenirs for sale to tourists visiting Kruger National Park, which was on their doorstep, but few of the tourists knew where they were sold. Ideas such as online marketplaces were discussed, where tourists might be able to order carvings or other souvenirs through a website they'd access from their hotel. Back then neither mobile phones nor Internet access were common in many of these areas, and there were many conversations like this about what technology *could* do. In many cases, little actually happened. As Richard and I discovered as we gathered evidence for our report, a lot was being said, and a lot of ideas were being thrown around, but actual projects with actual impacts were few and far between.

Our efforts with wildlive!, on the other hand, couldn't have been more contrasting. Here we were using the very latest mobile technology to build a cutting-edge mobile Internet site, and the Vodafone network had data connection speeds as fast as anything available at the time. Phones being sold in the UK had colour screens, played games and tunes, and could access limited Internet services. The phones we were seeing on our travels were far older, mostly Nokia models, with small, monochrome screens and functionality limited to sending and receiving phone calls and text messages (SMS). It was clear that any services anyone might develop, or propose, for users here would be very different from anything being developed for those back home. These contrasting technical landscapes would not be lost on me and, along with my wider interest in appropriate technologies, deeply influenced my work with FrontlineSMS, a project which would emerge a couple of years later.

Given all the buzz and excitement, you'd have thought mobile phones were a relatively new technology, but by the time I bought my Sharp GX10 they'd already been rolling off the production lines for about thirty years. Motorola started mass producing the first mobiles in 1973, but the idea came even earlier than that, in 1908 to be exact, when the first patent was issued for a wireless phone in Kentucky, USA. Ironically, for many years mobiles weren't really that mobile at all, and were often large and heavy, with awful battery life. It wasn't until the mid-1980s, when technical specifications for the GSM standard were approved (meaning all phones would be able to talk to one another, and work across borders), and the wider technology and computing sectors began their explosive growth, that things really started to take off.

But even then things took a while to get going. In the UK, mobile phone ownership was still as low as 15% when I first walked through the doors of Sussex University in 1996. Ten years later that would reach 80%, driven in large part by the launch of pay-as-you-go, which made getting a phone much quicker and easier.

Around this time, a little-known company seemingly came out of nowhere to dominate the market. Named after the Finnish town where they'd set up their second paper mill in 1868, Nokia was better-known early on in its history for producing paper, electric cabling, rubber tyres and rubber boots. It wasn't until the early 1990s that the company decided to turn its attention to mobile phones and mobile infrastructure, and the timing was perfect. By 1998 Nokia was the best selling mobile phone brand in the world, and one model in particular, the iconic Nokia 1100, ended up selling over 250 million units. Even today, despite their sad demise following Microsoft's disastrous 2013 takeover, Nokia still accounts for eight of the top ten best selling phones of all time. Before joining FFI, the only phones I'd ever owned were made by Nokia, and it almost felt like an act of betrayal buying anything else. No-one would have known, Nokia included, but we were just four years away from the launch of the first Apple iPhone, an event that would change everything, and signal the beginning of the end for them in particular.

Meanwhile, in Cambridge the summer came and went, and by the autumn we'd completed our scheduled Resource Africa trips. We had more than enough material to write what we hoped would be a useful report, and we felt well positioned, thanks to our field trips, to be able to separate hype and hope from truth and reality, and make a number of sensible recommendations to both the mobile industry and the conservation and development sectors. Our report, *Mobile Phones: An Appropriate Tool for Conservation and Development?*, needed to be ready for the printer by the end of the year, so Richard was kept busy doing what he did best, knocking it all into shape, while I returned to the wildlive! project.

As time went on I began to learn more and more about how mobile phones could, couldn't, should and shouldn't be applied to global conservation and development efforts. And as I went, I found myself being contacted by more and more non-profit organisations grappling with the kinds of challenges being thrown up in our research. Mobile phones were clearly generating plenty of excitement, but few people were actually doing anything with them, and as a result there were few projects to study and little best practice to speak of. And while many organisations saw some sort of potential for the technology, most simply didn't know where to begin, so most did nothing. Others tried and made a mess of it. I always felt in something of a privileged position sitting in my front row seat, and I increasingly felt a responsibility to share what I was learning as widely as possible, not just within the walls of FFI.

Driven by that obligation to share, late afternoon on 4th September 2003 I pulled out my credit card and registered 'kiwanja.net' as a website domain. I wasn't the only one beginning to take an interest in what was happening with mobiles, of course, but most other people were focused almost exclusively on the technology, registering names like Tech360 or Mobiles4Good for their projects. For me, what mattered first and foremost were the problems and the people and, if mobile technology could help, then great. But it was equally important not to make technology a fit when it wasn't. Whatever you named your website was, for me, an early signal of intent, an indication of how you planned to approach, and make the most of, the opportunity.

I'd stumbled across the word *kiwanja* a few months earlier, and liked its neutrality and sound. A Swahili word, it can mean anything from open space to field, to wasteland, to earth. kiwanja.net was to become my spiritual home on the web, and the place where, for the next 15 years, I would share my work, my ideas and my thoughts on how the conservation and development (and wider social innovation) sectors could do better with the opportunity technology presented them.

For me personally the year was shaping up well, but 2003 was also a big year for FFI, who were celebrating a significant milestone. Exactly one hundred years earlier, the Society for the Preservation of the Wild Fauna of the Empire was established by a group of aristocrats and American statesmen. Later renamed Fauna & Flora International, its founding objective was the preservation of large mammals in Southern Africa, populations of which were being decimated by hunting. By the 1900s, literally thousands of men and women were travelling to Southern, Eastern and Central Africa to shoot wildlife. One hunter reportedly managed over 1,000 elephant kills in his lifetime, and all manner of other big game including lions, rhinos and giraffes, were shot in huge numbers, their skulls, tusks, feet, horns or anything else deemed desirable shipped back to the hunter's home to display as trophies.

Arguments for the creation of national parks were central to the Society's efforts, and they pushed the British colonial government to create designated wildlife protected areas. The creation of parks in the Serengeti in Tanzania and Kruger in South Africa (where I would later work), as well as others in Kenya, have been put down to their lobbying efforts. For some people these were key accomplishments in African conservation, but for others the allocation of large tracts of land for wildlife which, in many cases meant the eviction of anyone living there, was nothing short of scandalous. We did, indeed, feel a certain irony working with communities in Bushbuckridge, some of whom had relatives kicked out of Kruger when it was created in 1926. It felt like FFI's work, and legacy, had come full circle.

Sir David Attenborough was Vice President of FFI during our time working on wildlive! (he still is, and FFI's new building in the centre of Cambridge was named after him). To celebrate the centenary it had been decided the year before to hold a black tie dinner at the Natural History Museum, under the imposing Diplodocus skeleton which proudly dominated the main hall. Tables would be sold to sponsors and supporters, and Sir David himself would attend. There would be a short film celebrating FFI's achievements, and a blind auction with some of the most incredible prizes I'd ever seen, including a gorilla safari and invites to private islands. As the development of wildlive! continued, and enthusiasm and excitement reached new heights, some bright spark decided that Sir David should officially launch wildlive! at the centenary dinner, too. In a stroke, things suddenly got serious. There was now no room for error. We'd have to be up and running by 11th December, else this would likely be my first, and last, major mobile phone project.

As we got closer and closer to the product launch we pulled in FFI's brilliant copywriter, Tim Knight, to help. Tim was freelance but had been involved with the charity for many years, and he'd authored a number of children's conservation books of his own. Not surprisingly he had a way with words, blended with a wonderful sense of humour, and we were soon drawing on his vast experience to help write copy for the website, and our launch materials.

Tim didn't disappoint. 'Satisfy your animal urges while you're on the move. Does a laughing zebra tickle your funny bone? Maybe a whooping gibbon is music to your

148

ears? Perhaps you prefer to be serenaded by a frog chorus? Whatever rings your bell, we've got a ringtone for you.'

And for the Silverback game, we settled on this: 'Put yourself in the hairy, oversized shoes of a silverback gorilla and pit your wits against his natural enemies. Can you feed yourself, stay healthy, find a mate and raise a family while dodging the poachers?'

While we knocked wildlive! into shape, Masabi continued to make progress with Silverback, and with increasing frequency we were being sent versions to share with staff for testing. Around the same time we were introduced to a full-blown commercial gaming company that agreed to adapt one of their games for us in return for a profit share, and this meant we'd have a couple of mobile games to offer at launch. And, to top it all off, the main website for wildlive! was also close to ready, and we were busy testing to make sure everything worked. By now I was working most evenings and weekends, doing my best to juggle multiple moving parts in an incredibly exciting but challenging project. This may have been work, but it rarely ever felt like it.

As December approached we began final preparations for the wildlive! launch, complicating yet further FFI's ambitious centenary celebration evening. Vodafone lent us around a dozen Sharp mobile phones, and we spent the best part of a day loading up all of our content onto them. We were told the signal inside the Natural History Museum was a good one, but decided not to take any chances. Our plan was to leave a phone on every table to give guests a chance to explore wildlive!, and a mixture of Vodafone and FFI staff would mingle in branded t-shirts ready to help answer any questions, Richard and I among them. None of us could have imagined, 11 months earlier as we all met for the first time in Cambridge, that our year would end with Sir David Attenborough launching the service we'd build, and at such a glittering event in such an incredible location.

Given the first ever meeting of the Society for the Preservation of the Wild Fauna of the Empire had taken place in the Natural History Museum, the choice of location could not have been more poignant. Known as the 'British Museum (Natural History)' until 1992, it took seven years to build, finally opening its doors in 1881. Around seven million people were visiting the Museum annually at the time of our launch, that figure rising rapidly to over eighteen million ten years later after the Labour government scrapped entrance fees. Today it employs over 850 people, and is the most visited natural history museum in Europe.

The Museum's mission to 'inspire a love of the natural world and unlock answers to the big issues facing humanity and the planet' was closely aligned with FFI's, another reason it made sense to celebrate the centenary there. Holding the dinner in the main hall, dominated by the skeleton of 'Dippy' and the most beautiful, bold lighting, gave the event an exceptional, almost eerie backdrop. Flower Valley Conservation Trust, a South African charity set up by FFI in 1999 to protect the unique flora of the Cape Floral Kingdom in the Western Cape, provided stunning fynbos floral displays for the tables. And, sitting among the flowers, silver cutlery, wine glasses and menus were our Sharp mobile phones.

The phones lent to us by Vodafone were the latest model, one up from the ones we'd bought earlier in the year. The newer GX20 had a great, vibrant little screen which brought our wallpaper images and games to life better than ever before. There was genuine curiosity among the guests, who seemed to enjoy listening to the range of animal ringtones, particularly after a few glasses of wine. After sharing the history of the project with Sir David, who was mingling between the tables to speak to the other guests, Karen turned and introduced him to the wildlive! team.

'Sir David, this is Ken and Richard. They've both been busy working on wildlive! for us.' We all shook hands, me trying my best not to come across as too starstruck.

'Hello Ken, Richard. I've been watching people play with their phones. That's quite an exciting service you've built there. Education, and awareness-raising, are so important for conservation.'

You can probably guess how much that meant coming from someone like Sir David. We chatted for a little longer before one of us said something that sent us both into fits of laughter. I have a photo, which hangs proudly on my wall, of me and Sir David standing between those tables, laughing hysterically. Who said what escapes me, but it was a lovely moment captured on a wonderful night, all wrapped up in an incredible year. I'd meet Sir David again, a few years later, when we'd share the stage at a conservation event in Bristol.

The evening, and the launch, were both a tremendous success. It was a genuine honour to have been there, and to be a part of it. Towards the end, as things slowly began to wind down, Richard and I took a short break and headed for a couple of wooden crates tucked away behind the stage. Beer in hand, we sat quietly for a moment, taking it all in.

'Cheers, buddy,' I said. 'I think that went pretty well.'

'It was brilliant, Ken.'

'It's been such great fun working with you, Richard. I just wanted to say that. Cheers to a year I'll not forget in a hurry.'

We raised our glasses, bringing them together with a reassuring chink. We allowed ourselves one brief moment of celebration, but we weren't done yet. The report Richard had been leading on was as good as complete, but it still needed a few finishing touches before it could head off to the printer in the new year. In January we'd also get news of the first month's income from the wildlive! service, a staggering £10,000. Silverback would also go on to receive rave reviews in the mobile gaming press, and wildlive! would continue to pull in valuable revenues for FFI for months to come.

Christmas that year gave me more than one reason to celebrate. Exactly twelve months earlier I had been grappling with my future in Jersey, self-doubt almost costing me the golden opportunity Karen had presented me. Although my contract was now officially over, more good news was on the way. The Vodafone Group Foundation had become big fans of our work with wildlive! and, in the new year, follow-on funding obtained by Karen would help take it to the next level with an expansion into other Vodafone territories across Europe. The Foundation also wanted us to explore some of the ideas we'd come up with in South Africa, and provided funding that would give rise to a completely new project which we'd undertake with our new friends at Resource Africa. Instead of facing the new year out of work, it turned out that my adventure would continue after all.

kiwanja.net was also waiting in the wings, a website I was yet to do much with but one that would play a significant role in my future career. And, to top it all, I finally felt my life had purpose. As the year ended, things really couldn't have been any better.

Interval

For me, 2004 will always be the year that drew the short straw, a relatively nondescript twelve months (compared to the others, at least) sandwiched between two of the most significant events in my life – the end of my lengthy search for purpose to one side, and the birth of a project which would take me to unimaginable levels on the other. That's not to say it wasn't a year without achievement, mind you.

Once again it started in Cambridge. This time, though, I didn't arrive at the FFI offices after the Christmas break with little to no idea what was to come. And this time I was living in an actual house with Elina, my Finnish girlfriend, and was no longer a paying guest in someone else's. Following a year of unknowns and uncertainty, now would be time to consolidate and confidently push on.

It started well enough. The first wildlive! income figures, which arrived by email around the middle of the month, totally blew us away. We really had no idea what to expect, no idea what a good month might look like. For FFI to receive over £10,000 of valuable, unrestricted income was beyond anything any of us thought possible. (Unrestricted income is gold dust in the charity sector because, unlike restricted funding it can be spent on anything, even the boring stuff, like rent.) It was a figure that was to repeat with surprising regularity over the coming year.

While the income flowed from our UK site, work progressed quickly as we translated wildlive! content into new languages, with our eyes set on launches in Germany, Hungary, Malta, Spain and the Netherlands. Although we'd been given the funding, we still needed Vodafone's buy-in before we could do any of the European technical integration. For that, Karen and I headed to The Strand in London, where their Director of Marketing for Europe was based. It felt like one of those *Eminem* 'Lose Yourself' moments. You know the one, where he sings about having one shot at something. 'Would you capture it, or just let it slip?' We'd have one opportunity to pitch wildlive! to Vodafone Europe's gatekeeper, and failure to inspire and excite him would see our ambitions dead in the water. Powerpoint presentation and GX20s to hand, Karen and I arrived at their glamorous Strand offices and, a short while later, left with smug, confident smiles on our faces. wildlive! and its stunning variety of unique content ended up selling itself.

A couple of months in and we were already on a roll. Things got even better when our *Mobile Phones: An Appropriate Tool for Conservation and Development?* report hit the shelves. It was the first time I'd ever seen my name on anything like that, and it was a huge moment for me. One of the first things I did, after staring at it for a while, was to send a copy to Jersey for my mum. I'd end up sharing all my successes with her, my way of saying thank you for all she'd done for me over the years. Without her none of it would have been possible, after all. I made sure she never forgot that.

By the spring we were ready to make our first trip to meet our Resource Africa colleagues. Some of our funding allowed us to dig a little deeper into the potential uses of technology in their work, and in wider conservation and development efforts across the continent. And it was there that we identified our first big opportunity.

Back then there was something of a disconnect between technologists and conservationists, and the two rarely if ever found themselves in the same room. There were no places online where they could meet and exchange ideas either, and we felt that if conservation was to benefit from all the tech innovation beginning to emerge, someone needed to put that right. Who better than us?

Looking back we were a little naive to think that you only had to put a bunch of techies in a room with a few conservationists for amazing things to happen. This was something of a prevailing view back then, and most people seemed to think that you just needed to create the right conditions and then fantastic new solutions and ideas would magically appear out of nowhere. Some people, even today, still believe this to be the case.

The argument that you only needed to provide people with Internet access, for example, then just step back and watch everything fall into place, was another popular myth at the time. It never 'just works out' of course, and when people did get online they rarely seemed to use their access to do the kinds of things the conservation and development communities wanted them to. I remember during my time in Nigeria, for example, visiting one of the new Internet cafes that had sprung up close to Ishie Lane. Glancing over people's shoulders, I could see that most of them were replying to spam email, trying to win cars, or looking for love – all fine, of course, but they were hardly accessing distance learning opportunities, building websites or apps, or looking up market prices for their goods, the kinds of things the ICT4D (Information and Communication Technologies for Development) community hoped they would do.

While our earlier South African trips had been dominated by lengthy drives out to rural communities, this year we'd spend most of our time trapped within the walls of office buildings. We remained hopeful – convinced, even – that providing an online space for conservationists and technologists to connect and share ideas was the best thing we could do, and the one piece that was missing. By the summer our plans for a new website, t4cd – Technologies for Conservation and Development – were well underway, and we got busy mobilising anyone and everyone we knew with an interest in, or working with, conservation and technology.

It was a strange mix of people to bring together, though. Most technologists were thrilled at the very suggestion that their solutions might be useful in helping wildlife conservation but few, if any of them, knew anything about wildlife or conservation. At the same time, most conservationists experienced a degree of fear when confronted with technology, seeing it as something that would likely only complicate and not aid their work. Many also worried about becoming dependent on something they knew so little about, and that might break in the sorts of challenging environments where most of them worked.

Concerns and mismatches aside, the idea that it was worth launching a new online community gathered pace. We hired a project manager, a Kenyan rugby fanatic by the name of Wallace, and planned a public launch for early 2005 at a special conservation technology event in Cambridge. We secured the services of Bill Thompson, a regular pundit for the BBC on all things technology, who would be our on-stage presenter and compere for the day. Bill would later play a significant role in the rise of FrontlineSMS, but that was still a couple of years away.

We also managed to convince Trevor Baylis, inventor of the wind up radio, to give a keynote speech. Trevor was something of a hero of mine, and someone I'd seen on television many times over the years. In real life he turned out to be a little shorter than I expected, but was a warm, welcoming character, always dressed in the same loose-fitting khaki jacket and trousers. He was never without his trusty pipe, either, and had a cheeky smile half-hidden under his bushy, grey moustache. He was one of the few people back then who had successfully reimagined and adapted a technology to work specifically in developing countries, making him an ideal motivational speaker for our event.

About 15 years earlier, Trevor had been sitting at home on the wonderfully named Eel Pie Island, a patch of land about nine acres in size that sits in the middle of the River Thames in London. While he sat watching television, a BBC documentary on HIV/AIDS in Africa caught his eye. During the first half of the 1990s, HIV infections were skyrocketing across sub-Saharan Africa and, with an estimated nine million cases out of a global total of around 14 million, the continent was having a torrid time. The international community was struggling to find ways of dealing with the pandemic, with treatment an unlikely prospect and awareness-raising campaigns aimed at prevention having little impact. It was the glaring need for mass messaging that made Trevor sit up. Despite leaving school at 15 with no qualifications, he had since become something of an accomplished inventor, and he wondered whether there was anything he could do to help.

The inability to get messages out far and wide was seriously impacting the efforts of the health authorities to spread awareness of the disease, and to let people know how they might avoid infection or be tested. Given their ubiquity, radios were widely considered to be the answer, but a lack of batteries, and the cost of replacements, prevented many people from listening to one. In less than an hour, using parts from a regular radio, a motor from a toy car and a winding mechanism from a music box, Trevor built the first prototype of what would become the wind up radio (later renamed the Freeplay radio). Ingeniously it had a hand crank and, with just a few turns it could generate enough energy to power the radio for up to 15 minutes at a time, making the need for batteries immediately redundant.

Trevor would later lose control of his invention, much to his disgust and frustration, and his talk at our event focused largely on how he was 'done over' and let down. An organisation he set up to help other inventors avoid falling into the same trap ended up helping thousands of people and launched numerous spin-off companies. Sadly, Trevor experienced financial difficulties in later life, and died on Eel Pie Island in 2018 at the age of 80 with over 250 inventions to his name. He was cremated in a coffin built in the shape of a wind up radio.

As we approached year-end, wildlive! was progressing well, and we were all set for the imminent launch of t4cd. Over the next twelve months it would become a fully funded project in its own right, one of the first, if not the first, attempts to build a community of interest around conservation technology. I was still loving everything I was doing, and was getting stuck into the kinds of projects that, only a couple of years earlier, I could have only dreamed of.

As I settled down to celebrate Christmas that year little did I know that, just a couple of weeks into the new year, a random idea would suddenly come out of nowhere, leading me on a journey that would change my life forever.

Hello World!

Ideas often have a habit of popping up in the most unexpected of places or at the most inconvenient of times. If you're lucky you might be somewhere you can write them down, to make a record of them somehow, before they slip back into the depths from where they came. Most of the world's ideas go that way, vanishing before they take hold, often gifted to someone unable to capture the moment, or ignored by those with no intention of acting upon them in the first place.

Of course, being blessed an idea is just the beginning. Without execution they remain at best fleeting glimpses of what might have been, at worst mere scribbles on paper. I still have dozens of ideas I'm yet to do anything with, scattered around in everything from cupboard draws to computer folders. I'll likely go to my grave before I get back to most of them. Thanks to the Coronavirus pandemic and a succession of lockdowns, this book is no longer among them.

Throughout my life I've always had a good imagination, and I'm often invited to join meetings or working groups because of my perceived creativity, or ability to think outside of the box. Perhaps I do have a talent, but conjuring up ideas on demand isn't usually one of them. My brain just doesn't seem to be wired that way. Rather, my ideas tend to come to me unexpectedly, such as on long walks, when I'm in the shower, while I'm cooking, lying asleep in bed, or simply watching TV. Whenever I've found myself stuck writing this book, for example, I've often managed to conjure up a way forward when I've been furthest away from the keyboard. Fortunately, most of the time I managed to make a mental note of what came to me, be it the answer to a tricky opening or closing to a chapter, or ideas for a new one altogether.

So it was by pure luck, then, that the one big idea I was to have in my life came to me while I was sitting comfortably at home, an arm's reach away from pen and paper, and not while I was standing in a gale on a remote hillside somewhere. Add the fact that African communication challenges were the furthest thing from my mind at the time, and my light bulb moment becomes even more of a mystery.

Indeed, that Saturday evening my thoughts were elsewhere as I lay on the sofa watching football on television, half asleep, with a near-empty bottle of Newcastle Brown Ale in my hand. Apologies in advance to anyone expecting a moment of inspiration on a par with Sir Isaac Newton, who had his a-ha moment as he pondered gravity while sitting under an apple tree. Or Archimedes who, taking a break from his study of mass, had his as he displaced water while getting into the bath. Newton and Archimedes were already investing huge amounts of time and brain power trying to understand those two particular problems, and it was very much on their minds when their big moments came. Their stories have since become a huge part of their legacy. Who doesn't think of apples when they think of Newton, or of Archimedes screaming 'Eureka!' as he jumped out of the bath before running naked down the street?

My moment couldn't have been more contrasting. Boring, even. I hadn't given a second thought to the communication challenges I'd seen as I travelled around Bushbuckridge a year earlier. Somewhere deep in the back of my mind I clearly recognised the problem, and possibly an opportunity, but I felt no motivation or obligation to do anything about it, and it had all become lost to the ravages of time. Why it suddenly came back to me that Saturday evening, at that moment, is something that confounds me to this day. But it did, and I'm eternally grateful for it. (It's a moment now immortalised in a short, animated National Geographic 'Live Curious' video, where a football bounces over a sofa with a bottle of beer precariously balanced on an armrest, as I explain the source of my 'big idea'.)

Within seconds I hastily put what was left of my beer down on the small coffee table next to me, and strode the short distance across the room to my desk, which sat tucked away in the far corner. Grabbing a pen and a scrap of paper I quickly jotted it all down. I didn't want to miss too much of the football, so I drew a quick diagram and labelled it up. A laptop computer, a cable and a mobile phone, and the odd scribbled word and an arrow here and there. That was basically it. I returned to the football, finished my beer, and within the hour was fast asleep in bed.

The following morning, after a lie-in and a quick, late breakfast, I sat down at my computer in my pajamas with a mug of freshly brewed coffee and decided to do a little research. I looked again at my rough notes from the night before. Was this new? Had it been tried before? If not, why not? Was it even possible? So many questions, and only one place to turn for answers – our old friend, the Internet.

It became immediately obvious there were plenty of platforms available that allowed you to send hundreds of text messages to people's phones from a computer, which is basically what I wanted to do. But every product I found needed three things that I knew would be a problem for organisations working in places like Bushbuckridge. Firstly, they relied on an Internet connection, something which was hugely problematic (it still is in many parts of the world, even today). Secondly, text message 'credits' needed to be purchased using a bank credit card, something else that none of the predominantly grassroots nonprofits or activists that I'd spoken to would have. Last but not least, because these products were all Internet-based, most were unable to receive replies from the phones they were messaging. You can't, after all, text the Internet – it doesn't have a mobile number. I wanted to create a genuinely two-way messaging solution, not something that was top down and one-way.

Remarkably, if my scrappily mapped out idea worked it would solve all of these problems, three problems which I believed, at the time, were the main stumbling blocks preventing text messaging from becoming a genuine tool for social change in developing regions of the world. After a couple of hours online I came to the conclusion that what I wanted to build didn't exist, not as I imagined it, anyway. Now I just needed to find out why. Perhaps it wasn't technically possible, in which case that would have been the end of it.

At the heart of my idea was a simple computer program that would run directly on a low-end, cheap laptop computer, one which would allow the adding of contact names and mobile numbers, and then the creation of groups to which those contacts could be added. The groups could then be selected and blasted with a single text message. The sender might want to inform community leaders in a particular village about the date and time of a proposed meeting, or share health information with a group of community health workers, or give market prices to groups of farmers to help them determine the best prices to charge for their crops. Once the message was typed into the computer, the user would select the group to send the message to and the system would send commands down an attached cable, to a regular mobile phone at the end, which would then send the messages out one-by-one over the local mobile phone network. I knew I could probably write the software to handle contacts and groups, but what I didn't know was whether you could use a cable to transmit a series of commands to a phone to tell it to send a text message, and if so what those commands were. I never shied away from a challenge, and this was the sort of challenge I loved.

It was time for a break, something to eat, a little fresh air and then one final bit of research. Everything would hinge on what I found out after lunch.

Years earlier, during my Clifton Computer days, I'd learnt how to use a piece of software called Carbon Copy. This nifty little program enabled me to connect with other people's computer systems remotely, over a dial-up modem and phone line, allowing me to carry out most technical support without having to leave my desk. Thanks to Carbon Copy I became familiar with the process of setting up and configuring modems, and learnt something known as the Hayes command set, a simple language that tells modems what to do and how to do it. These skills, picked up more than ten years earlier, would prove invaluable over the next few days as I set out to prove my idea.

Over the course of that morning things had fallen neatly into place, but I still didn't know whether a regular mobile phone, when connected to a computer with a cable, would act like the kind of modem I was used to, or whether it would act like a modem at all. And, if it did, whether any Hayes commands existed to tell it how to send and read a text message. I randomly selected a mid-range phone from the Nokia website, downloaded the technical manual, and began to read. Towards the back, hidden in an appendix, there they were, references to Hayes commands that could tell the phone not only how to send a text, but also how to read one back from the phone's inbox. My heart picked up a few beats. I felt so very close to proving the viability of my idea. All that remained was to try it out for myself, with real equipment.

I searched for a second-hand Nokia phone and the right cable to connect it to my Dell computer. This was early 2005, so I wasn't spoilt for choice with online shopping sites.

In fact, only eBay had what I needed, and fortunately I was already a member. After cross-checking available phones with the Nokia website – I needed to be sure that I not only bought one that supported Hayes commands, but one that would be available to users in developing countries – I bought a second-hand Nokia 6100, a CA-42 cable and a pay-as-you-go SIM card to use for testing. I hit the 'Pay' button, and logged out. Satisfied with my day's work I closed my computer, pushed all my notes to one side and got on with what was left of my weekend.

Over the course of the next few days the equipment gradually arrived. It was mental torture, the greatest puzzle of all arriving piece-by-piece, but by the end of the week I had everything I needed laid out neatly on my desk. My moment of truth was here. Everything would now hinge on what happened during the next few minutes.

I slid the SIM card into the Nokia phone, replaced the battery and back cover and powered it up. It had a couple of bars of power remaining, just enough to do the tests I needed. I connected the CA-42 cable to the base, waiting for the reassuring click, and screwed the other end into the 9-pin serial port at the back of my computer (these were the days before everything came with more familiar USB connections). I logged into Windows and opened Terminal, a small Microsoft Windows program that allows users to communicate with any devices physically connected to the machine they're using. The cable was plugged into the only communications port I had on my Dell, so I knew the phone sat at the end of COM1. Using a combination of my own memory, commands I found online and notes from the Nokia technical guide, I changed a couple of default settings and, after a minute or two, was ready to go.

The first thing I needed to do was to see whether the phone could be 'seen' by Terminal. If not, nothing would happen no matter what commands I sent it. The quickest way to do this was to send the briefest of commands – the letters 'AT' in fact – down the wire and wait to see if the device responded. In simplest terms, it's the Hayes equivalent of 'is anybody there?' If something is there, and it understands that it's been asked to say hello, it will respond with a short but reassuring 'OK'.

I held my breath as I typed 'AT' into Terminal, before hitting the Enter key on the keyboard. Immediately, without a second's hesitation, the phone responded. There it was, two bold letters that said it all. 'OK' appeared on the computer screen. My breath quickened a little more. The phone was there, and the computer could not only see it, but could talk to it as well. Just one step remained. Could I instruct the phone to send a text message using commands I'd type into Terminal, and then read one back from the phone onto the computer screen? This was the biggest and most important test of all, and the further I got the more disappointed I knew I'd be if I fell at the final hurdle.

Using the Nokia reference manual, I looked up the command to tell the phone what number I wanted to send the text message to, which happened to be my Sharp GX20, and sent it down to the Nokia phone. Next, I followed it up with another command containing the actual text I wanted to send in the message, 'Hello World!' (it's coding tradition to use 'Hello World!' as an opening message the first time you try something like this). Then it was time for the final command, the one to tell the phone to send the message. By now the anticipation was killing me. Breath well and truly held, I hit the Enter key one last time.

The blinking cursor on the computer came back with 'OK', followed by a pause, a pause which felt like an eternity. I stared nervously at the GX20 screen, waiting for any kind of response. It was all or nothing. Had I done everything right? Was the message going to arrive? Was I just wasting my time?

It turned out I wasn't. The message arrived, and I couldn't contain my excitement. The reassuring series of beeps from the Sharp told me it had worked, and I read the message back. 'Hello World!' indeed. I hit the reply button on the phone, typed a short message and hit the send button. A few seconds later the Nokia, still sitting on my desk with its cable attached, beeped reassuringly. I could now see my reply on its little monochrome screen. This was fantastic! But now the big question. Could I get my Dell computer to read it? I dived back into the technical guide and, a couple of Terminal commands later, and a little more holding of breath, the message appeared in plain text on the computer screen.

I couldn't believe it. I was over the moon. Ecstatic. It felt like some sort of magic trick to have been able to send and receive a text message without having to physically touch a phone, and to do it all via a series of commands on a computer keyboard. Less than a week after dreaming up the initial idea, I'd done everything I needed to prove it was, indeed, possible.

I sat down with a cup of coffee and, despite the caffeine rush, managed to come slowly back down to earth. I couldn't help but imagine all the things I could get a system like this to do, and all the people in all the places it might help. All that was left now was to write the software to run on the computer, software which would drive everything for the user. For this to work they'd need to be shielded from all the technical stuff happening in the background, and all the commands and configuring I'd just done manually. This was an exciting proposition, although I'd not done any coding of any real significance since leaving Jersey for university. I'd never written a Windows application before, either. I didn't know it yet, but there was going to be a steep learning curve ahead of me.

Before I went much further I decided to share what I was tentatively calling 'Project SMS' with Karen and Simon, who I was now working with on a few smaller, independent projects. I typed out a one-page concept note explaining what I hoped to build, what it could do and where it might be useful, and highlighted the differences between what I was planning and what already existed. I included a hastily cobbled-together image of a laptop with a phone and a cable attached to help them picture what the whole thing might look like. Karen was still in touch with her contacts at Vodafone, and we approached one in a personal capacity for help. If I was going to develop this piece of software then I would need a laptop of my own to write it on, a copy of Microsoft Visual BASIC to write it with, a few technical guides and a couple of different test phones and GSM modems on which to test it. We added a bit extra to the budget to cover some of my time, and settled on £10,000 which we figured would be enough to get Project SMS off the ground.

Things moved quickly, and in more ways than one. Simon did what Simon does best and, knowing I wasn't happy with my makeshift project name suggested we call it 'FrontlineSMS' (Simon was the one who came up with the name wildlive!, too). The new name was a perfect fit, and immediately stuck. The FrontlineSMS domain was bought a little later in March, ready for a website that we'd need to build when we eventually launched. The small amount of seed funding we were hoping for came through, and soon I was out shopping for a laptop, researching phones and modems and taking my first tentative steps into the world of Windows programming. It would be another few months before coding would begin in earnest, though. I did, after all, have other paid work on at the time and didn't want to launch myself into FrontlineSMS until I'd done all the necessary research, laid all the groundwork, and was able to focus my attention fully on the task ahead. My apprehension was understandable. I wasn't entirely sure how things were going to work out, if they were going to work out at all. Would I even be able to write the thing?

FrontlineSMS development was pushed to the back of my mind over the next few weeks and months, but as I went about my daily business I continued to seek out opportunities for learning and for partnerships, or to join random dots I thought might help me, come summer, to hit the ground running. I became something of an expert at opportunistically dropping FrontlineSMS into random conversations and, one morning, my approach paid off in spectacular fashion.

That March I was invited to an event at Microsoft's main campus in Cambridge where I did what I was increasingly asked to do, and spoke about the potential social benefits of mobile technology. During the break I bumped into one of the product managers responsible for Visual BASIC. It was a moment of pure serendipity. Andrew was friendly and approachable, and the fact he liked my earlier work, and my talk, certainly helped motivate him to help with what I was planning next. At that time I hadn't even started exploring Visual BASIC, the programming environment I'd decided to use for FrontlineSMS, but I knew it was likely I'd need expert help, most likely at crucial moments. Cheekily, I asked for his help, and remarkably Andrew offered to make himself available and share any questions I had with the Visual BASIC developer team. We exchanged email addresses and, true to his word, he and his team were brilliant when coding finally began that summer.

Days and weeks passed by and the summer months were soon upon us. Elina, who was by this time studying full time at university in Sussex, had been offered a job with a family friend back home in Finland throughout August. Over a drink one warm, sunny evening in the beer garden at the Prince of Wales, our village pub, we talked through our options. Elina leaned forward towards the shady half of the table.

'So, Jussi has offered me some work over the summer, and I'm wondering if we could both go and turn some of it into a holiday.' She hadn't been home for a couple of years and was keen to go back to catch up with family and friends.

'Well, I need the time, and a load of headspace, to tackle this new FrontlineSMS project, so that could work', I replied. I knew Forssa well enough, and knew it would be quiet and relaxing, allowing me to focus on my coding. 'How long are you thinking?' I asked.

'Maybe five or six weeks? University is closed so I have as much time as you can get, to be honest.'

'Okay, let me check tomorrow. If August looks clear, we should just go ahead and book the tickets', I answered with a smile, empty glass in hand. 'Another drink?'

It was a no-brainer, to be honest. Elina needed the money, and I needed time and space to lock myself away and tackle FrontlineSMS, so heading to her family home in Forssa for a few weeks felt like a win-win. I would have the house to myself as everyone would be out working all day. We went ahead as planned, and a few weeks later we arrived in Finland, back to the town I'd left in frustration exactly four years earlier.

Laptop, manuals, cables, phones, modems (and, more importantly, coffee) to hand, I set up shop on an old mahogany kitchen table which sat in the back corner of their lounge next to the window, giving me a wonderful view of the fields and forests outside. The room was typical of most Finnish houses, with an open-plan living space leading into the kitchen, all decked out with wooden flooring. Pushed up against the walls was the most basic of furniture, with plain wallpaper acting as a backdrop to scattered photos of family members – past and present – which were dotted around in the sort of old fashioned picture frames you usually only find in antique shops or museums. On the wall above the sofa hung a large picture of a lion resting under a tree, painted by Chiti, a welcome and constant reminder of my time in Zambia. A little further along sat an old, neglected, out-of-tune piano which Elina used to play occasionally as a child. The lounge, basic as it was, turned out to be an ideal place to work and, shortly after we arrived, I settled down into what would become a five-week coding marathon. It's hard to believe that when I sat at that old table for the first time I had no idea what I was taking on. By the end I'd have exceeded all my expectations, and would get on the plane home with not only a solid version of FrontlineSMS to share with the world, but also a fully functioning website.

The coding was, as expected, a challenge, but the kind of challenge I enjoyed. I was now living in the world of Microsoft Windows, and I had to think about things in an entirely different way to what I was used to. In BASIC programming of old, you always knew which line of code the computer was running, and you'd simply give the system instructions about what to do next, a decision usually based on what the user entered on the keyboard. If, for example, you were just waiting for something to be pressed, the system would simply sit at that line of code and wait. Visualising your code running was a relatively straightforward and simple thing to do.

Windows turned out to be a different beast altogether. When a user sits in front of a typical Windows application, there are usually a number of things they can choose to do at any one time. They might click on Cancel to go back, or click inside a text box to type something, or scroll down a list and select an entry, or click on a drop-down menu, or drag something somewhere. Basically, after laying out the screen your program needs to just sit and wait and, depending on what the user decides to do, different bits of code in many different places might be called to run. Because of this, Windows is known as 'event driven', and it was quite different from anything I had used or coded in before.

In all, I had to design about a dozen screens for FrontlineSMS, thinking carefully about each one and then deciding where on the screen each of the boxes, buttons and lists should go. This was the most enjoyable part. Dropping different elements onto a page and sizing and moving them around until they all looked right wasn't that difficult, and I always fancied myself as a bit of a designer. But that's where the fun ended. I also needed to think about how I'd 'string' all these different screens together, how I would pass information between them, and the specific function each had to carry out. And I had to do it in a way that would be immediately obvious to the users, many of whom would have little experience with computers, or messaging software come to that.

Remarkably, I had no detailed notes, no plans, no diagrams, no list of functions. I broke all the rules and just built it as I went, relying on intuition and a combination of gut feeling and luck. Three years later I would be interviewed by a Nokia engineer, and he would be disappointed to learn that I hadn't followed any established 'design principles' in my work. I didn't even know there were any. It was seen as something of a miracle that I'd built something as effective as FrontlineSMS without, on the surface at least, really knowing what I was doing.

Every time I got stuck and couldn't find the solution in one of my hefty Visual BASIC manuals, I'd turn to the Stack Overflow website, an online community of developers who seem, at some point, to have answered every conceivable question about every conceivable computing problem in history. If, by some miracle, there wasn't anything useful there I'd down tools and go for a leisurely stroll through the endless forest that lay on my doorstep. Aided by the warm sunshine and gentle breeze, these walks did wonders to clear my mind. They were spiritually refreshing, too, a great way to escape and think about anything other than text messaging and coding. Putting distance between me and my keyboard usually did the trick, and I'd often return to my laptop and dive straight back in with a new lease of life.

When it didn't work I'd reach out to Andrew at Microsoft, my knight in shining armour. Usually within a day I'd have a reply, forwarded on from one of the team who provided the sort of high-level technical support that, without Andrew, I could have only dreamed of. I would reconnect with him many years later, during one of the pandemic lockdowns, and let him know how incredibly valuable his contribution to the FrontlineSMS story was. It may have been nothing to him, but it meant the world to me.

The more I progressed and battled my way through each major coding obstacle, the more my confidence grew. Just a couple of weeks in and things were already beginning to take shape, and I was really enjoying writing software again. I even managed to relax a little, allowing myself the occasional daytime distraction.

Summer that year coincided with one of the biggest sporting rivalries in the cricketing calendar – the Ashes series between England and Australia – and radio commentary was streamed on the BBC website. Both sides would duel over a small brown urn which, according to legend, contains the ashes of a wooden bail, burnt after a match in 1882 when Australia beat England on English soil for the first time. The Ashes represented the 'cremated body of English cricket' and the sides have fought over the tiny four-inch urn ever since. That summer the series was in England, and I listened online to the daily commentary as I merrily coded away. It was great company and, at times, a great distraction, particularly when Australian wickets fell. Occasionally I sent messages to the BBC Cricket commentary team, and twice they were published on the BBC website. The messages are still there, a nice reminder of a summer I will always look back on with great fondness and pride.

One morning in late August I settled down as usual with my coffee when an email popped up in my inbox. It was from a journalist at the Charity Times who was writing an article about how non-profits were using mobile technology. She asked if she could send over a few questions about wildlive! and my work in South Africa. I agreed and, opportunistic as ever, in my reply I managed to squeeze in a mention for FrontlineSMS, even though it wasn't finished yet.

A couple of days later my phone rang.

'Oh, hi Ken. We exchanged emails earlier in the week about this Charity Times article, and I wanted to quickly check something with you, if that's okay?'

'Sure. Happy to help. Fire away.'

'So, I'm drafting up a paragraph on FrontlineSMS. Can you remind me when you hope to have it ready?'

'I'm aiming for October,' I replied, confidently.

'Okay, great. And one last thing, where can people go to find out more?'

It was then that it dawned on me. Website. I had no website! Where would people go to get more information, or to sign up to try it out? I'd just been presented my first big marketing opportunity, and I was about to blow it.

'Oh, that would be www.frontlinesms.com.'

'Great. Thanks, Ken. I'll let you know when the article goes live, hopefully in the next week.'

I hung up. Plans for further coding were immediately put on hold as I frantically put together the first FrontlineSMS website. I had all the content, albeit in my head, so I just needed to add some structure to it, add a few images here and there and get it live. One important thing was missing, though – a banner image for the top of the home page. Short of ideas, I quickly took a photo of the field and forest outside my window, and by sheer luck it was a perfect fit. It still brings a smile to my face whenever I see archived versions of that very first website, a reminder of my view each day as I coded away. By today's standards the website is nothing to write home about, but I don't think it was bad at all for 2005. At least now I had one and, as it turned out, it did the job perfectly.

The combination of coding and bouts of walking, relaxing and drinking Finnish beer on the banks of some of the many local freshwater lakes, all in the warm, endless daylight, turned my trip into the perfect working holiday, and progress with FrontlineSMS was swift. Five weeks always felt incredibly ambitious to write a messaging platform from scratch, especially with no plan, but as September arrived it was as good as complete, and I'd managed to throw in a website for good measure, too.

Crucially, because I'd written everything myself, it would be simple to fix anything that didn't work, or add anything I may have missed, without the need to hire anyone to do it. Being able to support the ongoing technical development of the platform, as well as provide technical support to many of the early users, turned out to be crucial to its early survival and later success. It would be two years, during which time I ran it all as a hobby, before any significant funding would come my way. The importance of keeping things lean, particularly at the beginning, turned out to be one of the key bits of advice I would share with other would-be founders and innovators for years to come. You'll find it mentioned later in this book, too.

We returned home to our one-bedroom flat during the first week of September and took a couple of days off before Elina headed back to university to prepare for her second year of study. I would then be free to get back to my consultancy work, adding any finishing touches to FrontlineSMS during the odd evening and weekend. As it crept ever closer to completion, I began thinking about how I'd get word out. I tentatively contacted various groups and journalist friends who I thought might either be interested in trying it, or be able to help share news of the launch. FFI would have the honour of hosting the first ever public demonstration when I planned a lunch time talk for 29th September, an event that felt particularly poignant given it was my work there that had inspired the idea in the first place.

As I continued to think about exciting ways of promoting FrontlineSMS, I stumbled across a BBC news report about a brilliant new website launched by a 21-year old British student by the name of Alex Tew. It was something I quickly decided FrontlineSMS needed to be a part of. Alex was short of money to cover his university fees, and came up with the idea of building a web page exactly one million pixels in size, selling each pixel (or dot on the screen) for a dollar. On 26th August 2005, the Million Dollar Homepage was born and it quickly captured the Internet's imagination. By October it was already well on the way to filling up, so I waited no longer and took the plunge, purchasing 200 pixels for FrontlineSMS. It was the only money I would spend promoting the platform in the early years and, although the Million Dollar Homepage might not have been the best strategic fit, I knew it would become part of Internet folklore. The last block of one thousand pixels was put up on eBay in January 2006, selling for $38,000 and signalling the end of Alex's incredible adventure. The website is still online today and, hidden among the hundreds of images, you'll find two small squares carrying the graphic of a tiny mobile phone, an image that provides one of the earliest links to what was then the first ever FrontlineSMS website.

By now I was only three weeks away from my own launch, and the coding for FrontlineSMS should have been done and dusted. But this was no ordinary project, and I guess I was no ordinary programmer. As a sign of the organic nature of it all, an idea came to me randomly one afternoon as I sat on the top deck of a London bus. As it stood, FrontlineSMS allowed you to add contacts and put them into groups, and to send and receive messages to and from those groups, and it did that brilliantly. What was lacking was some sort of automation, something interactive. And that's what came to me on the bus that day. What if I added an 'AutoReply' feature, I thought to myself. Users could text in a specific keyword, and then FrontlineSMS could automatically respond with a predefined message, anything from a bit of health advice, or the opening hours of a clinic, or the price of cassava. It sounded like something fun to try and write but, more to the point, it sounded pretty useful.

I immediately realised the value of adding this sort of functionality, so that evening I got to work on adding what, this time, really would be the last big change. I was done by the weekend, and it worked a treat. I could now leave FrontlineSMS sitting unattended on my desk, send in a text message from anywhere and have it respond without any human intervention. I didn't realise it at the time, but this one piece of automation was to prove incredibly powerful when increasing numbers of community groups began taking an interest in the platform over the coming years.

October 31st, 2005. The big day. With a mixture of excitement and anticipation, I pulled up a chair and sat down at my Dell computer, typing out an email to all my contacts, friends and supporters with the subject 'The FrontlineSMS system is now ready for trial'. Eyes fixed on the screen, I took a deep breath, allowed myself a wry, contented smile, and hit 'Send'. A random idea that had come to me in that very same room, just ten months earlier, was now a reality. There was no going back, it was out in the world, and it was now up to the world to decide what would happen next, if anything.

Around the time of the launch I was in the early stages of a three-month contract with TTPCom, a Cambridge-based technology company that hired me to help test their mobile phone operating systems. I'd found the job randomly, casually glancing in the window of one of those recruitment agencies in Cambridge city centre a couple of weeks earlier. It was fun, and it paid well. The TTPCom engineers, who sat a few metres away, would do their usual software development and testing, and then hand everything over to me and two colleagues. Our job was to run through a series of tests to make sure everything worked as expected, and if it did we'd then move on to trying out random things to check how stable the code was. Because of my own programming background, I became particularly good at figuring out ways of crashing the phones, a success rate that did little for my popularity among my coding colleagues.

While they scratched their heads trying to figure out what I'd done and how I'd done it, there was little else for me to be getting on with. So I decided to take advantage of this unexpected free time and turned my attention to promoting FrontlineSMS. I'd search for anyone and everyone online who was involved with technology in Africa, or mobile phones, or activism, or humanitarian response. I'd email journalists, academics, nonprofits and technologists and share the launch of the platform, asking if they wanted to give it a go, or if they could help spread the word, offering myself up for interview in the process. There really wasn't anything like it at the time, and the origin of the idea in Southern Africa and the simple, offline grassroots focus made it a relatively easy sell at a time of growing interest in the use of mobile phones in global development.

The first person to break news of the launch was Mike Grenville. Mike ran one of the best and most popular SMS-focused websites at the time, 160 Characters. Having him share the news was a big deal, and Mike and I would later become friends. I was also asked to write a guest post for MobileActive, a relatively new and rapidly growing community of activists and practitioners interested in the potential of mobile phones for social change. These two articles, more than any, gave my outreach efforts a real boost, pushing the news into precisely the kinds of places I was hoping for.

One of those places was Zimbabwe. Reading the MobileActive piece from her office in central Harare was Brenda, co-founder of one of the country's most prominent civil society organisations, Kubatana. Against all the odds, Bren and her co-founder, Bev, had been actively and very publicly promoting human rights in the country, despite the obvious risks they took working right under the nose of the Mugabe regime. Kubatana had been something of a pioneer in the use of email to share news and information with fellow citizens, but they recognised the obvious limitations given Zimbabwe's weak technical infrastructure, poor Internet penetration, and low computer ownership. They had been taking more and more of an interest in mobile phones, and SMS in particular, as a way of spreading their message further.

Zimbabwe is often associated with the endless turmoil and violence of President Robert Mugabe's 30-year grip on power, but it wasn't always that way. Geographically, Zimbabwe is a beautiful, landlocked country, sharing its borders with South Africa, Botswana, Zambia and Mozambique. It was first colonised in the 1890s when white settlers arrived, and the country remained largely under colonial rule until 1964 when Ian Smith became Prime Minister. His first act was to demand independence from Britain and, when this was denied, he unilaterally declared it anyway, causing international outrage. At the time, Rhodesia (as it was known) was home to just 200,000 whites compared to four million blacks. Smith believed that black majority rule would never happen 'in a thousand years'. How wrong he'd turn out to be.

A certain Robert Mugabe had become increasingly active politically during the 1960s, a period of rising African nationalism, and as leader of the Zimbabwe African National Union (ZANU) he played a key role in the struggle for independence, a struggle that would ultimately lead to the deaths of over 30,000 people. Despite the loss of life the struggle eventually paid off and, in 1980, the country was finally granted independence from Britain. Robert Mugabe was declared Prime Minister. Seven years later he would change the constitution and become President, a position he would hold with an increasingly iron grip for the next 30 years.

At the time of independence, Zimbabwe (as it was now called) held great promise. Often referred to as the 'Jewel of Africa' it boasted an abundance of wildlife, and was one of the few African countries to be home to all of 'The Big Five'. And let's not forget Victoria Falls, one of the seven wonders of the world. The soil was rich and the country was a net exporter of food. Earlier modernisation meant they had good roads and pavements, solid infrastructure, a number of airports and an education system that was the envy of the continent. As independence dawned on Friday 18th April, 1980, the future looked bright.

How things would change. Zimbabwe's promising start slowly eroded away, to be replaced by years of corruption and mismanagement. An economic crash followed, leading to a crisis of faith in Mugabe's leadership. Despite controlling much of the machinery of government he suffered a humiliating defeat when, in 1999, his proposed constitutional reforms were roundly beaten in a national referendum. To make matters worse a new party, the Movement for Democratic Change, or MDC, was emerging as a rare and very real threat to his hold on power. When black farmers started intruding on, and eventually seizing, white-owned farms as part of a growing dispute over land, Mugabe saw it as a way of regaining the initiative. Not only did he do nothing to stop it, he actively encouraged it. International donors pulled funding in protest, leading to a decline in foreign currency reserves, food shortages and further economic collapse. As if that weren't enough, Mugabe then restricted freedom of the press, and ramped up his blame on Britain and the international community for Zimbabwe's woes. Despite the chaos, Mugabe's party claimed victory in the March 2005 elections with two-thirds of the votes. The MDC cried foul. That summer, tens of thousands of shanty houses were raised to the ground as part of a government 'clean up operation' that, not unsurprisingly, targeted opposition MDC strongholds.

This was the scene when Kubatana reached out to me on 15th November asking about FrontlineSMS. Having followed Zimbabwe's gradual collapse – it was a staple on BBC News at the time – I was delighted to hear that my fledgling platform resonated with an organisation doing such amazing work in one of the most challenging of places. I immediately warmed to Bren and Bev, and worked with them over the next couple of weeks to get them up and running. In no time at all they were using FrontlineSMS not just to send news updates around the country, but to collect news from communities at the sharp end of the violence. I remember thinking to myself at the time that if Kubatana were the only people to use FrontlineSMS, I'd be happy.

As it turned out, there would be many, many more.

Into the Wild

The next couple of years were productive, if not spectacular. FrontlineSMS didn't exactly set the world on fire, but it was exciting to see it slowly picked up by a good number of organisations doing all kinds of exciting work in all sorts of faraway places.

In Aceh, Indonesia, UNDP and Mercy Corps began using it to send market prices and other agricultural data to smallholder rural coffee farmers. In Iraq it was used by the country's first independent news agency to disseminate news to eight countries, and in Afghanistan it helped keep fieldworkers safe through SMS-based security alerts. In Malawi it started generating interest in the health sector where a project started by Josh Nesbit – a Stanford University student – used it to set up a rural healthcare network for 250,000 people. It was used by bloggers in Pakistan during the 2007 state of emergency to get news safely out of the country, and in the Azerbaijani elections to help mobilise the youth vote. In Kenya it helped report breakages in fences caused by elephants, and ran the Overseas Filipino Workers emergency helpline that ensured workers received immediate assistance in case of personal emergency. It was deployed in the DRC along with Ushahidi, a crowdsourcing platform created by friends in Kenya, to collect reports of violence around the country, and helped the Grameen Technology Centre in Uganda communicate with their rural customers. And a network of journalists used it to help report and monitor elections in Ghana, Guinea and the Ivory Coast. The list went on.

Most of this was to come, though, and there was still plenty of hard work ahead before I'd get anywhere near where I wanted to be. Luckily for me I was doing something that I not only loved, but that I turned out to be pretty good at – evangelising the platform, providing technical support to the users, generating buzz and excitement in the social and technology sectors, and enthusiastically sharing the stories of those whose work it was helping. And because it all remained a simple hobby for the first couple of years,

I was under no pressure to work to arbitrary targets or deadlines, or to follow any pre-agreed plan with donors. This turned out to be a blessing. Over the years I've seen too many projects put themselves under unnecessary pressure early on, many failing not because they didn't have good ideas, but because they ran out of money after being encouraged to build organisations, rent offices and hire teams before they knew whether their idea was useful, worked, or was even wanted by anyone.

When I wasn't busy with my day job (you know, the one that paid the bills) I continued with my 'scattergun' approach to promotion, reaching out to anyone and everyone who I thought might have the slightest interest in FrontlineSMS. Sometimes it paid off, sometimes it didn't, but I knew the simple act of getting the name into people's inboxes might plant a seed that would later lead to greater things, even if nothing happened straight away. Playing the long game came naturally to me, and luckily I wasn't in any kind of hurry.

Despite my own lack of urgency, some opportunities did move rather quickly, though, and one email exchange in particular kicked off a conversation that would end up taking me all the way to California.

As I sat in my modest, one-bedroom flat in rural Cambridge, sitting in his office on the other side of the world, in far grander surroundings, was a young Swedish innovator by the name of Erik Sundelof. Erik was a Fellow on the Reuters Digital Vision Program at Stanford University and he, too, was working on a project using mobile phones. The Digital Vision Program selected around a dozen individuals each year from all over the world and invited them to Stanford to work on projects that leveraged technology for good. Funded by Reuters, the global media giant, the Fellowship was a prestigious affair and I felt an immense sense of pride when Erik reached out to me by email on 7th December, enquiring about FrontlineSMS.

Erik turned out to be a rather tall, solidly built chap with wavy, strawberry-blonde hair, just the hint of a Swedish accent, and a passion for cycling. An ever-present, brown leather satchel hung over his shoulders, with a large MacBook Pro usually poking out of the top. He always seemed to have a smile on his face, and was the kind of cheerful character you'd willingly spend time with down the pub. He would turn out to be one of the nicest, most thoughtful people I would ever meet, and I remain forever grateful for his friendship and support during the very early days of my own career. Not only was he kind but he was also incredibly smart, and a born collaborator, and the Reuters Digital Vision Program was lucky to have him.

Erik had similar interests to me, and was spending his year at Stanford working on a system he hoped would make it easier for citizen journalists to file reports through their mobile phones, particularly in conflict zones. 'In The Field Online' was the first of its kind, allowing a combination of text, audio and video to be sent from a mobile directly to a website. It was our shared focus on building simple, appropriate tools that worked in the hands of local organisations and everyday citizens that drew him to my work with FrontlineSMS.

A couple of weeks passed after our first email exchange, and we decided it was probably worth jumping on a call. We settled on early evening in the UK, which was mid-morning for him. As I gradually found myself spending more and more time on international calls, I began to appreciate the challenges that these time differences would make. Calls to the West Coast of the USA were particularly challenging given the eight hour gap, but I was happy to speak any time of day or night, grateful for any kind of interest in my work.

'Hey, Erik. How are you doing? It's good to be in touch. I've been reading about 'In the Field Online' and it all looks really interesting.'

'Thanks. I've only just heard about FrontlineSMS and it piqued my interest.'

'Great to hear that, Erik. I'm happy that the news made it all the way to a Swede living in California', I replied, smuggly.

'I'm glad it did, too, and it's great to connect with people doing similar work to me. You know, I doubt I'm going to actually use FrontlineSMS, but I did want to learn more about why you wrote it, and how it's all going.'

We spoke for the best part of an hour. It was a remarkably easy conversation for two people who barely knew each other. By the time we neared the end we'd somehow stumbled our way into a much wider range of subjects, including the ice hockey rivalry between Sweden and Finland and what we thought of *Coldplay*, the British band riding high in the charts at the time. It wasn't until we were about to hang up that Erik dropped one final, throwaway comment.

'You know, Ken, you should apply to the programme here.'

'Oh, you think?' I replied.

'Absolutely. You're doing exactly the kind of work we're all focused on, and you have a great technical background.'

'Well, California does look like a great place to live, and getting to Stanford would be something special.'

'Why don't you dig around the website and I'll have a chat with Stuart, who runs the programme. I'll see what he thinks and let you know. But I think you'd be perfect.'

It wasn't lost on me that Stanford was one of the most prestigious universities in the world, and the top fundraising institution in the US (it was the first US college to raise an eye-watering one billion dollars in a single year). Stanford boasted a dizzying array of famous former students, a list which included numerous astronauts, presidents, Nobel laureates and several dozen billionaires. It was part of Silicon Valley folklore thanks to its unparalleled success in attracting startup funding, and many of today's largest and most successful tech companies have their own founding stories tied up in some shape or form with the university. Architecturally it was like nothing I'd seen before, boasting incredible sandstone buildings with huge, imposing arches, hugged by beautifully manicured lawns, all spectacularly wrapped up in a brilliant blue sky and the warmth of never-ending Californian sunshine. Stanford had so much going for it, it was little surprise so many people fell over themselves to get there.

Life carried on as normal for the next few weeks. I continued with my consultancy work and promoted FrontlineSMS during evenings and weekends, tweaking the software if anyone requested any changes or found any bugs. The idea of applying to Stanford didn't completely go away, although I was convinced I didn't have the slightest chance of getting in. The Reuters Program only accepted a handful of Fellows each year, and there would almost certainly be hundreds of applicants. Only about 10% of regular students were successful in getting a place at the university back then. For someone with a generally lousy record of achievement, not to mention a shaky academic record, my doubts were certainly justified. I had to remind myself that I had turned a corner, though, and things had been on the up over the last couple of years. Maybe this time things would be different.

March soon arrived and, after much thought, I decided I had nothing to lose by sending in a proposal to Stuart. I felt my technology background, experience working in developing countries, work on wildlive!, the mobile phone research Richard and I undertook and now the steadily growing FrontlineSMS project all added up to a very solid application. Erik certainly seemed to think so. I wish I'd been as confident as him.

My instinct proved to be right.

'17 May, 2005
Dear Ken I Banks
Thank you for your application to the Digital Vision
Fellowship. Unfortunately, we are unable to offer you a
Fellowship in the 2006-07 Program.'

Here we go again. After the disappointment had sunk in I emailed Erik to share the news. (I didn't burn anything this time, in case you were wondering.) I thanked him for his time, and said that I did think it would have been too good to be true. He was disappointed for me, too, which was a measure of the kind of thoughtful person he was. Apparently competition for Fellowships had been fierce, with over 200 people from 79 countries applying for only a dozen places. So, you see, the odds were always heavily stacked against me. That didn't stop it from hurting, though.

After a couple of days I still hadn't come to terms with the outcome, which I naturally disagreed with. Heaven knows I'd failed plenty of times before, so I should have been used to it. But I'd always found some way of picking myself up and either moving on, or battling just that little bit harder to bend the outcome in my favour. I knew that getting to Stanford would be transformational, personally and professionally. Refusing to give in just yet, I wondered if it might be worth trying a different approach. Rather than apply as a regular Fellow focused on my own project, how about I join and use all my skills and experience to support the work of the other Fellows? I could code, I'd built technology solutions before, I'd consulted, I had experience in mobile, I'd run businesses and I had experience working in the sorts of places many of the Fellows would be working. I couldn't help but wonder if that might be more appealing to Stuart.

On 25th May, just over a week after the initial rejection email, the crazy idea for a Collaboration Fellowship was born.

The rest, as they say, is history. Stuart warmed immediately to the idea and, to his credit, he did an about turn. It took a while to get the confirmation email – there was a Bank Holiday weekend in the way that drew out my agony for an extra 24 hours – but it did arrive, and it brought news that I'd made it onto the Program. Some things are worth fighting for, after all. Stanford would give my credibility a huge boost, and would open doors that would never have otherwise opened. That Fellowship would become the very first piece of news I'd publicly share on the kiwanja.net website.

I now found myself juggling relocation plans with FrontlineSMS, and FrontlineSMS with paid work. I already had an exciting few months ahead, but things were about to get crazier. Kubatana, who by now had settled well into FrontlineSMS, were working on a platform of their own. Known in those early days as 'Dialup Radio' (then later, 'FreedomFone') it was based on IVR, or Interactive Voice Response, where people spoke (or pressed number keys on their phones) at key stages of a pre-recorded call to determine what information they wanted. Kubatana saw IVR as yet another way of getting much-needed information out to citizens across the country who lacked Internet access but who had basic phones. To make the system as accessible as possible, and to help those with low airtime balances, people would be allowed to send in a text message and Dialup Radio would call them back, bearing the cost of the call. For this to work they needed to include software to collect text messages from a bank of GSM modems, and then pass them to an online service which made the calls back. Much to my excitement, I was asked to help them build it. And not just that, to go to Zimbabwe to do it.

First, I'd need to go shopping. In their efforts to control the flow of digital information around the country, the Mugabe regime had heavily restricted the supply and use of GSM modems. They weren't available for sale within Zimbabwe, and it was difficult and risky to ship them in. Our only option was for me to buy them, and then bury them deep in my luggage. We decided that, if I was stopped by customs, I'd plead ignorance and claim to not know what they were. The worst thing that could happen, I thought, would be a night in confinement before being thrown on a flight back home the next day. I wasn't as worried as I probably should have been. My mum, on the other hand, thought I was crazy. Zimbabwe was not a place to be messed with.

On 4th July I departed Heathrow for Harare. I had been to Zimbabwe before of course, in 1993, but that was only for a few minutes after that short walk across the bridge spanning Victoria Falls. This time was for real, and if I'm honest it was something of a dream come true. I'd met many Zimbabweans working in conservation over the years, and they all spoke very fondly of their country. Now would be my chance to see it for myself, and to be part of an exciting project working with some brilliant people. All that remained, after I'd left the comfort of the aircraft cabin, was to get in.

'Arrivals card!' the Zimbabwean customs officer barked, menacingly.

The man now blocking my way, staring me straight in the eye, was tall, a little rounded, sweating in his immaculate black uniform and cap, and not looking like the friendliest of characters.

'Here you are', I stuttered, handing it over and trying hard not to look as nervous as I felt. He snatched it out of my hand, without removing his gaze for one second. Talk about intimidating.

'What brings you to Harare?' he asked.

'I'm here on holiday, seeing some friends', I replied. He looked at my arrivals slip and grunted, before stepping to the side and waving me through.

I thanked my lucky stars that I got through customs unscathed that day. It turned out that they were more interested in stopping Zimbabweans whom they suspected of bringing overly-generous (and illegal) quantities of alcohol, cigarettes and other items back for their families from abroad. As those ahead of me were searched, hassled or aggressively questioned, I'd grown increasingly nervous, expecting to join them. With Mugabe's non-stop anti-British rhetoric it was a miracle I wasn't pulled aside and made to suffer, even just a little.

Bev and Bren met me outside the airport and, almost immediately, we hit it off. They were a tremendous, smart, lovely couple who complemented each other perfectly. Calm and gentle, well spoken, thoughtful and considered, their personalities were somewhat at odds with the violence and chaos ripping Zimbabwe apart. While this was primarily a work trip – Kubatana had used valuable funding to get me there – I sensed I was going to get a lot out of it personally, and that it would create memories and foster friendships that would last a lifetime.

It didn't take long to drive the ten miles from Robert Gabriel Mugabe International Airport to the Bronte Hotel in Harare. The hotel was only a short walk from Kubatana's offices so, for convenience, it was decided I'd stay there for the first week. After that I'd join Bev and Bren on their annual holiday which, this year, would be wild camping in Mana Pools National Park. And, finally, for the last week or so I'd be a guest at Bren's sister's cottage up until my departure at the end of July.

The Bronte Hotel was just shy of a hundred years old. A pretty, Cape Dutch style building, it was decked out in pure white and was rather colonial-looking, surrounded by beautifully manicured lawns bordered by stunning Jacaranda trees, with fine Shona sculptures dotted around the grounds. The sweet scent of flowers and the gentle buzzing of contented, pollinating bees filled the warm air. The climate was pleasant, nothing like the suffocating humidity I'd experienced in Nigeria. Things were eerily quiet and calm compared with the hustle and bustle of daily life just a short distance outside. The hotel service was as good as anything I'd had anywhere, as was the food, all of which sat rather uncomfortably with me given the hardship most people were facing in the rest of the country.

My fellow guests and I weren't completely sheltered from the realities of life outside, though. The impact of Zimbabwe's rampant inflation was increasingly reflected in the prices we were paying for our meals. Each time I ate at the Bronte, the conversation quickly turned to the subject of money.

'Good evening, Mr Ken. How are things?'
'Not too bad thanks, Tendai. How are you doing?'
'Yes, good thank you. So, not everything on the menu is available today, and the prices have changed again, so before you order anything please check with me first.'
'Of course. I had spaghetti carbonara the other day. Do you have that?'
'We do, but that's gone up to 400,000 dollars I'm afraid.'

Just a couple of days earlier I'd paid 150,000 Zimbabwean Dollars for that meal. As a visitor I was able to take whatever price hikes were thrown at me, my American currency surprisingly picking up in value the longer my trip went on. This all sat very uncomfortably with me. I did my best to help the staff at the hotel, tipping generously in US Dollars, but the economy was in freefall, and there was little I could do to make any real difference. Most Zimbabweans looked on helplessly as the money in their pockets, or their life savings if they had any, lost value at breakneck speed. By now, I'd noticed that few of the surviving shops in the city bothered to put price labels on anything anymore. Prices simply changed too quickly. Cash machines emptied in minutes, and the sight of bank notes in the gutters said it all. I left the country with a 50 billion dollar bank note in my pocket. It's now a bookmark, and my children stare at it in amazement.

On a happier note, my first few days in Zimbabwe coincided with the 2006 Football World Cup which, that year, was being held in Germany. In total, 32 teams would battle it out for the FIFA World Cup Trophy over four action-packed weeks, culminating in the main event on 9th July. That year the final was between France and Italy, with Italy winning on penalties after extra time. Impartial to the overall result, Bev, Bren and I watched the final together on television over a few bottles of local beer. It wasn't lost on me that this was the third World Cup Final in a row I'd watched on TV in an African country. In 1998 it was in Uganda. In 2002, Nigeria. And now, in 2006, I was watching from Zimbabwe. I did wonder whether it was worth trying to keep the tradition going, but come 2010 I'd given up on the idea. I can't even remember where I was that year, but it was almost certainly somewhere far less glamorous, and certainly not anywhere near Africa.

During my daily walk from the Bronte to the Kubatana office I'd regularly pass by Herbert Chitepo Avenue, a relatively short highway running in a straight line, east to west, through Harare. During my time with Resource Africa I'd worked closely with a wonderful colleague by the name of Kule Chitepo, and it had been casually dropped into the odd conversation back then that his father had been a significant figure in Zimbabwe's struggle for independence. Kule never spoke about it himself, mind you. Until I walked along a short stretch of it that first day I wasn't even aware there was a road named after him, and I didn't appreciate the hugely significant role Kule's father had played in the birth of his country.

Following a spell studying in South Africa, and then the United Kingdom, Herbert Wiltshire Pfumaindini Chitepo returned home to Southern Rhodesia in 1954 as the first black African to have qualified as a barrister. Shortly afterwards he established his own law firm and practised as a lawyer, defending a number of African nationalists in court. He was well known internationally and became leader of the newly formed Zimbabwe African National Union (ZANU). In March 1975, at the height of the struggles, Chitepo was assassinated in Lusaka, Zambia, a car bomb, placed inside a wheel arch of his Volkswagen, killing him and his bodyguard outright. The case remains unsolved to this day. ZANU infighting was one of the theories put forward, although it's considered more likely that the Rhodesian security services planned and carried out the attack. Following the assasination, a certain Robert Mugabe took over the reins at ZANU. Today, Chitepo's body lies in Heroes' Acre, a national monument in Harare. National Hero Status is the highest honour that can be bestowed on an individual by the Zimbabwean State.

My first week with Kubatana was dominated by coding, just as I'd expected it to be. If I wasn't ankle-deep in Visual BASIC manuals I'd be borrowing chunks of code from FrontlineSMS, or getting tangled up in cables, modems and SIM cards. Despite irregular power supply and a feeling that I was being followed everywhere I went, by the end of the first week we had a fully working version of DialUp Radio's new SMS feature. My hacked-together FrontlineSMS-inspired code would poll six modems, one at a time, passing any incoming text messages to the web server hosting the rest of the IVR system. From there the phone calls would be made. Everything worked better than we could have hoped and, by the end of the week, we were in a perfect place to take a much-needed break.

Bev had been busy the previous couple of days planning our stay in Mana Pools, cooking frantically and packing enough delicious pre-cooked meals to last us all week. She did an incredible job, and we wanted for nothing. With everything packed and ready, we headed off late on the Saturday morning. The roughly six-hour drive to Mana Pools was as interesting and revealing as the destination itself would turn out to be, with abandoned farm after abandoned farm dominating the landscape. It was a sorry sight for a country that had once boasted one of the most productive agricultural and tobacco sectors in the whole of Africa.

Mana Pools National Park is like no other in Africa. Yes, it's incredibly beautiful and is home to a vast diversity of wildlife, but so are many other parks across the continent. What makes it different are its walking safaris. Mana Pools is the only national park that allows visitors to roam freely, unguided if they wish. There are no walking trails, and no predetermined routes. Visitors are free to go anywhere and everywhere. Fortunately for us, Bren, Bev and I were accompanied by Richard, a calm, confident, reassuring, no-nonsense former soldier with a lifelong interest in the bush, in particular the Zambezi River valley. He would be our guide. Spending a week outdoors in the wild, camped right on the edge of the Zambezi River, with the Zambian border a short but risky swim away was incredible, and I'm forever grateful for being given the chance to experience it. Richard was incredible, too, taking us on winding walks through the bush, spotting movement and signs of wildlife that would easily have passed the rest of us by – or got us killed.

Camping in the bush carries obvious risks, of course, and it's a place that commands the utmost respect. A German tourist had died just a couple of weeks before we arrived after finding herself caught between a female elephant and her calf. The elephant panicked, charged, and that was that. The stretch of river where we camped was heavily populated by hippos, too, an animal responsible for more deaths across Africa than almost any other animal, about 3,000 per year by some estimates. From my own experience, hippos don't tend to walk around minor obstacles such as tents. They just walk straight through them. We needed to have our wits about us, day and night.

My biggest concern, though, were the lions we heard during our first evening. Sure, the roars were quite far off in the distance at the start, but with each passing night they seemed to get nearer, and louder. By the end of the week they sounded way too close for comfort. Sitting alone in my tent, in the pitch black, I felt in need of a little reassurance.

'Richard? You awake?' I whispered.

'Sure am. What is it?'

'These lions. They sound pretty close tonight.'

'They are. Probably less than a hundred metres away right now, by my guessing.'

'So, errrr ... Are we safe?'

'I think so.'

'Oh. I'm not sure I like the sound of that. Can you be more specific?'

'Well, it's unlikely they'd go for a tent, as long as we're calm and quiet.'

By now I was sitting bolt upright, my senses working overtime, alert to every sound and movement outside. I tried my best to reassure myself that Richard knew what he was talking about but, feeling increasingly vulnerable, I reached out for my hiking stick which lay on the ground beside me. The thought of fighting off a lion with a thin piece of wood brought a wry smile to my face, but that quickly disappeared when the next roar came. It felt so ridiculously close. Tens of metres away.

I'd never heard a wild lion before, or any lion this close-up come to that. It sounded nothing like the lions on television, or in a zoo. Survival mode kicked in, and in my heightened sense of awareness I could hear every single deep, throaty gasp of air as each loud, intimidating roar slowly wound back down to a quiet, steady breath. I felt so ridiculously vulnerable. We might be the smartest and most dominant species on the planet, I remember thinking to myself, and we may have invented the most incredible technologies, but at that moment my fate lay in the hands, or giant paws, of a four-foot high 400-pound killing machine. If one of those lions had decided my time was up there was literally nothing I could have done about it. Experiencing that level of vulnerability is quite humbling, and not something you ever forget.

Richard's curiosity finally got the better of him, and I heard him slowly unzip his tent.

'Let's all jump in the car, Ken, and see if we can spot them.'

'I'd probably feel a little safer doing that, Richard,' I replied, still holding firmly onto my stick.

My initial enthusiasm waned when I realised the car was parked a few metres away, turning my dash into more of a sprint. But after the reassuring clunk of the car doors, we'd all made it. We drove around for a few minutes, zigzagging between the trees and bushes with our headlights on full beam, piercing through the darkness.

'Anyone see anything?' asked Richard, hopefully.

'Nothing', I replied, peering anxiously through the window. Bev and Bren couldn't spot anything, either. To our relief, the lions appeared to have passed through, and our clash with one of the big five was over.

Incredibly, our most spectacular brush with nature was yet to come. We'd already got up close and personal with a small herd of elephants on one of our afternoon hikes, and that was exhilarating enough. And then, of course, there were the lions. But it was something that happened during our final morning, an incident that lasted all of 30 seconds, that topped it for me. It was dawn, and the gently warming sun was slowly bringing the day to life. As we sat having breakfast, the waters of the Zambezi gently flowing by, a gazelle flew out of nowhere, ripping right through the middle of our camp. It didn't care one bit about us and, given how panicked it looked, it was clearly running for its life. Before any of us could say or do anything, a pack of African wild dogs shot past in hot pursuit. A couple followed the line of the gazelle, running right past us, while the others took a wide berth as they attempted to encircle it. I'd foolishly decided to leave my Canon camera at home but I did manage, very quickly, to snap a grainy image of one of the dogs with a cheap disposable camera that I'd bought from Boots in Heathrow Airport. Not surprisingly, the photo was rubbish and did nothing to capture the excitement, and rarity, of the moment.

Also known as painted wolves or hunting dogs, African wild dogs are extremely elusive, and to say we were incredibly lucky to see them would be a gross understatement. Conservationists regularly spend months in the bush hoping for a glimpse, only to leave disappointed. While they're no threat to humans, they're one of Africa's most successful hunters thanks to their incredible intelligence and teamwork. Packs usually consist of around ten individuals, but groups of up to 40 have been sighted, and in full flight they can hit speeds of up to 40 miles an hour. I have no idea what happened to the gazelle that morning, but the odds were very much stacked against it.

Sadly, African wild dogs are critically endangered, threatened mostly by hunting and habitat fragmentation. Zimbabwe is one of just seven countries across the whole of Southern Africa to have viable populations, and today only 6,000 of these magnificent creatures still roam the continent.

Without a shadow of doubt, my time at Mana Pools was one of the most rewarding experiences of my life. The location was stunning, the wildlife rich, the food delicious and the conversation stimulating. And it was calm and peaceful, as far removed from Zimbabwe's troubles as you could get. A week was never going to be enough, but it was a week I'll never forget. After one final hike that morning, we slowly began to pack up, ready for the six-hour drive back to Harare and a return to civilization, with all its challenges and conflict.

My last few days were spent in a lovely little cottage in a large gated compound, owned by Bren's sister. We would leave for the office each morning and spend the days in very much the same way we did during the first week, building out the radio platform and making sure everything was working as well as possible before I left. I still managed to carve out a little time to explore Harare, heading into the city centre to get one last taste of the place, and to seek out mementos of my trip.

One stop I was determined to make was the legendary Book Cafe, a place I'd read about years earlier. Opened in 1993, the same year I first visited the African continent, it became one of Zimbabwe's most popular and vibrant locations for political discussion and cultural celebration. Over the years it's been described as a networking space, a music venue, an 'embassy of change' and a place for free-thinkers to gather and share ideas over food, drink and live music. Sadly, I could only get there late one Saturday morning and it was empty, so I missed the chance to sample the atmosphere of the place. But I could imagine what it might have been like, the haze of cigarette and cigar smoke drifting through the dim lighting, the chinking of bottles, the shuffling of feet and the groove of a live band, all battling with an orchestra of voices holding the liveliest of conversations. In its heyday, Book Cafe would have been quite a remarkable place to spend an evening.

Although I could still wander through it, the cafe itself was closed, but I did manage to buy a couple of books from the small bookshop to the side, books written and published by Zimbabwean authors that you had little chance of finding anywhere else. These were the sorts of books I loved to read, local voices that were otherwise absent from much of the political debate I got at home. There were few PhDs or global development experts here, just voices with lived experience, voices so often ignored but ones that counted for so much.

My time in this wonderful, yet deeply divided and troubled country, had to come to an end. With hugs and kisses all round, I bid farewell to Bev and Bren and left my Zimbabwean adventure behind me, heading home for what remained of the summer. We all kept in touch, and continued to work together and support each other. I'd meet Bren again a few years later in Victoria Station on one of her trips to the UK, where we shared lunch and a beer and reminisced about old times, as old friends often do.

I never did see Bev again. We sadly lost her in 2018, a loss which I still find very hard to come to terms with.

California Dreaming

'Next!'

After what felt like an age, it was finally my turn. I stepped forward and placed my passport down on the counter in front of me. I looked straight ahead and smiled, anxiously. I enjoyed travelling, but always hated this bit. The immigration officer stared back, expressionless, flicking his way through to the photo page of my passport before glancing back up. I didn't know if he was tired, bored, lacked motivation or simply enjoyed the power trip. His gun, can of pepper spray and set of handcuffs, all neatly arranged and on full display around his belt, did nothing to calm my nerves.

'So, what brings you to San Francisco, errrr ... Mr Banks?'

'I'm here on a year-long Fellowship at Stanford University', I replied, trying not to sound too nervous.

He thumbed through my passport for a second time, instinctively stopping on the page holding the stamp for my American visa.

'Can I see your paperwork, sir?'

'Oh, what paperwork is that?' It hardly seemed possible, but my nerves went up a notch as I took a long, hard gulp. We'd only got as far as his second question, and I was already faltering.

'The paperwork you filled in to get your visa, that would have been checked, stamped and given back to you at the embassy. That paperwork.'

That paperwork was in my suitcase, which by now was probably making it's umpteenth lonely lap of the conveyor belt in the arrivals hall. It dawned on me that I'd probably made a big mistake here.

'I'm sorry, but I didn't realise I'd need that. I thought the visa in my passport was all you needed to see. The rest of the paperwork is in my suitcase I'm afraid.'

'In that case, sir, I'm going to have to ask you to come with me.'

Before I knew it I was being escorted down a long corridor to a small, stuffy immigration office at the back. I wasn't in trouble, not yet, anyway. A few other people in the room looked as though they already were. This was the Naughty Room, and I was in it. After describing my suitcase as best I could, I was told to sit tight while one of the officers sighed and grudgingly headed off in search of it. This wasn't the welcome I had been hoping for, I have to be honest.

After what felt like an age my bag was finally rescued from its conveyor belt hell, and so was my paperwork. I promised to make sure I carried it with me in my hand luggage in future. With a mild slap on the wrist and a lengthy delay, I was finally on my way, out through arrivals into the blazing sunshine to meet my ride. 'Sorry I took so long,' I said, apologetically. 'I had a bit of a problem with immigration.' With that, we headed off to my new home, a rented room in a neat little house in Los Altos, a short drive down the road from Stanford campus. I was now officially in the country and could relax. The weather was beautiful, the air was fresh, everyone looked healthy and happy, and everything seemed clean and new. This was California exactly as I had hoped, and imagined, it to be.

In my haste to secure somewhere to live I'd settled on accommodation a little further away from campus than I should. It turned out to be too far to walk, which would have been my preferred method of travel. I had no intention of buying a car, so the only remaining option was to get a bike. My new landlady took me out shopping for one that first evening, and I was all set for my first trip to Stanford in the morning, a 30-minute ride along the long, nondescript Foothill Expressway. The journey might have looked straightforward enough, but it didn't stop me from getting lost that first day. Stanford is a huge maze of a place, and finding my way to Cordura Hall, where the Digital Vision Program was based, was far from easy for a first-timer with little sense of direction and even fewer points of reference. I did make it, and pretty much on time, taking comfort in the fact that most of the other Fellows also struggled with their directions that first morning.

There was plenty of free time built into the timetable, as you'd expect for a programme like this. Whenever we had scheduled get-togethers their main objective was to share our progress, critique each other's ideas, and help plan a way forward. We did have the odd lecture, though, and these would cover everything from AI and machine learning to business models or pitching to investors. Although my primary role was to support the other Fellows, I made sure I took in as much as possible in the hope that it might, someday, be useful in my own work with FrontlineSMS.

The first couple of weeks passed in a flash. We all settled in well, and began to socialise almost immediately – no surprise, really, given we were all new to Stanford and none of us knew anyone there. The Rose & Crown Pub would become a regular haunt, ably supported by Erik who stuck around after his Fellowship had ended. He knew all the good places to eat and drink, and the Crown was right at the top of University Avenue just outside Stanford campus, and perfectly located opposite the Cheesecake Factory, one of the best places to eat, even if it wasn't the cheapest. Despite the landlord's best efforts to recreate the atmosphere of a genuine English pub, the baseball and American football memorabilia scattered all over was an obvious, and humorous, point of failure.

I'd only been cycling in from Los Altos for a couple of weeks but was already beginning to realise the error of my ways. Although it wasn't a huge distance, it made getting to campus and back more challenging than it needed to be, and the crazily hot Californian weather meant that I rarely arrived in the best possible shape. I was also finding myself missing out on many of the sporadically-organised social activities. To top it all, renting a room was far from cheap anywhere in or around Silicon Valley, and I wasn't looking forward to watching my life savings dwindle away over the next 12 months. I knew something would have to give, and soon.

Fortunately there was an alternative. I've always been a fan of simple living, and I've always loved camping, sleeping on floors, or roughing it a little. I find it a leveller of sorts, a good way of grounding myself, a reminder of what matters. I'd also enjoyed the freedom of moving around and living out of suitcases the past few years, and given I was now in California I wondered whether I had it in me to live out of a van. It would solve my rent problem, and if I could park up close to Stanford it would solve my distance problem, too. (Not many people lived out of vans back then, but this has changed in recent years as rents in the area have become totally unaffordable.)

Most of my Stanford colleagues thought I was crazy, but I dived into Craigslist and immediately found a nice 1973 VW Westfalia for sale down in Long Beach. It was going for a pretty reasonable $6,900 and was described as being in 'excellent condition for the year'. If this had been the UK then a vehicle of that age would have been close to falling apart, I thought to myself. The van had been in the same family from new, another selling point, with the son having inherited it from his parents when they bought a replacement some years earlier.

I contacted the owner and the van was still for sale but, not surprisingly, there had been a lot of interest. Long Beach was about 400 miles away, so arranging to 'pop down' and see it made little sense for a whole bunch of reasons, and it was likely to sell before I could get that sort of trip organised. I asked the owner, Rick, if he could hold it for me, and we agreed that I'd pay a $500 deposit and promise to get there within three days. If I liked it I'd pay the balance there and then, and drive it back. If I didn't like it, or didn't show up, I'd lose my deposit. Rick was convinced I'd fall in love with it, though, and as it turned out he'd be right.

Despite my excitement I didn't fancy a 14-hour, 800-mile round trip on my own. I'd never driven in the US before and had only been in the country a few weeks, so my apprehension was understandable. Fortunately one of the Fellows from the previous year (who, like Erik, had decided to stick around) decided a road trip would be a pretty neat thing to do. Atif was always up for an adventure and I was grateful for his enthusiasm, and his company. We made plans to drive down the coast on the Sunday and see the van in the evening. All being well we'd stay overnight in a motel close by, get the cash to pay for it in the morning, and then drive it back, assuming it turned out to be as good as Rick claimed it was.

Late morning on Sunday, 8th October, we picked up our rental car from the Enterprise depot on the edge of campus, bought some snacks and drinks from one of the nearby shops, and set off. Not surprisingly, the drive was a dream. What's not to like about a road trip with good company in a large, comfortable, air-conditioned, good-as-new automatic car, with wide open roads the whole way, blessed with a healthy dose of Californian blue sky and sunshine? Our only challenge would come later that evening when we booked into a roadside motel, one of us having to sneak in to avoid paying for two people. Luckily it all worked out perfectly, even if it might have looked slightly dodgy to anyone outside catching a glimpse of what we were up to.

Evening arrived, and the light was already beginning to fade as we neared our rendezvous with Rick, an outdoor shopping centre car park on the outskirts of town. I may not have been your typical hippy VW camper owner, but Rick certainly was. Well built and tanned, with light brown, wavy hair, baggy shorts and a ripped t-shirt, he spoke with the calm Californian drawl I was already getting used to. He was so laid back he was almost falling over, as the joke goes. I could imagine him pulling up on a golden, sandy beach somewhere, leaping out of the side of the van and sprinting down to the sea, a surfboard strung firmly and confidently under his arm.

Atif and I approached the camper, with Rick waving at us enthusiastically through an open window. At first glance it was exactly as he had described, and it looked in pretty amazing condition for a near 25-year old vehicle. The side door slid open and he jumped out athletically.

'Hey, Ken! It is Ken, isn't it?'

'Sure is. Great to meet you, Rick', I said, nodding. 'This is Atif, a friend of mine. He was pretty excited at the prospect of coming down to get the van.'

'Hey, man', replied Rick, reaching out to shake hands. 'Well, here she is, Ken. As you can see, she's in pretty good nick. Tyres are clean, bodywork's excellent, engine runs well. You wanna come inside?'

The three of us stepped up, one at a time, through the sliding side door. Atif and I took a seat at the back, and Rick spun the front passenger seat around to face us. It did look great inside and I immediately warmed to it. 'Never been smoked in', said Rick, proudly. The back seat doubled up as the bed, and there was a small wardrobe in the corner behind us. The sink was to the left, along with a two-ring gas cooker and a small shelf. I could imagine myself living in here, my own little private world, my own little bit of freedom, everything I owned packed into this cosy space. It was exactly what I wanted, and needed. Despite parting with $500 before I'd even seen it, and then spending a couple of days and another couple of hundred dollars getting there, it seemed my gamble might just have paid off. Some things just feel like they're meant to be. After a little more discussion, everything that needed to be said had been said.

'I'll take it, Rick. It's perfect.'

'Awesome, man. I knew you'd like it! There are only a few early Westies around in this condition. You've got yourself a good one.'

And so, with a firm shake of the hand the deal was done. I didn't bother haggling on the price. I'd not have to rent an expensive room in a house any more, and the van would pay for itself by the time I left Stanford. That was good enough for me. We arranged to meet outside a local Wells Fargo Bank the following morning where I'd withdraw the remaining $6,400. We'd sign the paperwork and then be done. Atif and I headed off in the rental car to grab a bite to eat, and then find a place to stay for the night, more than happy with our day's work.

'What do you think, Atif? Of the van?' I asked, as we pulled out of the car park.
'I love it, Ken. It looks perfect. I'm a little envious to be honest.'
'Well, you've been a massive help, Atif. Dinner's on me tonight, okay?'

The following morning we met Rick outside the bank as agreed. We went through the paperwork one last time, and then I strolled excitedly over to the Wells Fargo entrance. I pushed the door. Nothing. Locked. I checked the time on the 'Opening Hours' sign. Strange. It should be open. Why wasn't it? I looked through the window at the empty hall inside, puzzled. Catching the reflection of a gentleman walking past on the pavement behind me, I turned around.

'Excuse me. Do you know when the bank is going to open?'

'Oh, you're out of luck today', he replied. 'It's Columbus Day. All the banks are shut.'

Columbus Day? What on earth was that? I gingerly walked back to the camper to find Rick and Atif deep in conversation. Rick glanced up. 'So, how did it go?' I explained what had just happened. Somehow it had escaped us all that it was a Bank Holiday. You could argue that Atif and I had an excuse not knowing, but Rick had simply forgotten. There was nothing for it. We'd have to hang around another day and try again the following morning, our adventure destined to go on for just that little bit longer.

By the craziest of coincidences, that Monday was Rick's birthday. In typical laid-back fashion he hadn't planned anything, so we decided we might as well all hang out for the day. Universal Studios was a short drive down the highway and, with little else to do we headed there, celebrating with a few fairground rides and the odd cold beer. This road trip was getting weirder and weirder by the minute, but in a good way. For Atif it was an adventure that just kept on giving.

The next morning we returned to the bank, and this time I got lucky, leaving with a brown envelope stuffed with cash. After carefully counting it out and handing it over, we said our goodbyes, Rick giving his beloved camper one final glance and a gentle, friendly tap on the roof. With that, Rick was gone. Atif and I dropped off our rental car and were finally on our way. We had the best part of a day to get back to Stanford, which was a relief given we'd never driven an old van like this before, and it was understandably a little temperamental. As we got closer to campus we pulled in for a break at a local shopping centre, and while we were there I took the chance to start kitting the van out. Pillows, duvets, coat hangers and towels were top of my list. Later I'd buy a radio to listen to NPR, and get myself a portable DVD player. Often, when I wasn't working, I'd lie in the darkness watching documentaries, or listening to the radio late into the night.

By early evening we'd made it safely back to Stanford campus, and I dropped Atif off. I was grateful beyond words that he'd come along. Not only did his companionship ease my nerves, but he was a lovely guy and great fun to be with. It had been a longer-than-planned trip but one I'll never forget, for all sorts of reasons, and without doubt he played a big part in it. After saying our goodbyes I drove the short distance along Foothill Expressway to Los Altos, where I was still living, and parked the camper a short distance up the road from the house. That evening I told my landlady I would be moving out. She was disappointed, and I felt bad, too, given I'd promised to stay longer when I first took the room. She thought I was crazy attempting to live in a van, and predicted it would end in failure – as did most people to be honest – but I was determined to make it work and didn't see it as a hardship at all. It was an adventure within an adventure as far as I was concerned, and probably the only chance I'd ever get to experience a way of life I'd wanted to try out for years. If you can't live in a van in California, I thought to myself, then you probably can't live in a van anywhere.

As I saw out my notice in the Los Altos house I made arrangements to get a parking permit for the van. Stanford had three parking zones, with the one in the centre the most expensive, and then the second and third becoming progressively cheaper as you moved further and further away from the heart of campus. Fortunately for me I would be in a 'C' zone, which was not only the cheapest but was also closest to Cordura Hall where the Digital Vision Program was based.

The parking space I would lay claim to was in the top corner of the car park, and about as far away from the centre of campus as you could get. Few parking attendants bothered walking that far most of the time, so I was able to park up my van, facing Sand Hill Road – preferred location for many of the valley's richest venture capital firms – for the whole of my Fellowship without attracting any unwanted attention. In reality, vehicles weren't meant to stay in the same car park beyond a couple of days at a time, let alone in the same space, and living in a van on campus was hardly encouraged. But I only ever got one parking ticket, right at the very end of my stay. Living simply, in a van, turned out to be one of the best things I did, something that kept me grounded as I worked in one of the most incredibly beautiful and well-off places imaginable. It's all too easy to get lost in the comfort and beauty, and forget why you went there in the first place, and that was something I was determined would not happen to me.

The Digital Vision Program progressed well, and in between project time and structured seminars we all made trips to various companies in and around Silicon Valley to learn more about what they did, and how they might help us. I was enjoying my role as I got to know each of the Fellows, supporting them in any way I could. It kept life varied and interesting. Whenever I had the chance I'd continue to push FrontlineSMS but, to be honest, that was slowly becoming sidelined as newer interests took hold. I started to work on other ideas, including a donations platform for grassroots nonprofits in developing countries. FrontlineSMS remained interesting but, with total downloads still under one hundred, it hadn't hit anywhere near its potential despite my efforts, and I wasn't sure where it was going, if anywhere.

Even if the software itself wasn't getting the traction I hoped for, it was doing a great job of opening doors for me. First, the International Telecommunications Union (ITU) invited me to become a member of their Connect the World initiative. Then, a few weeks into my Fellowship, I was contacted by Stephane Boyera, Chair of the W3C Working Group on the Mobile Web in Developing Countries, who was organising a conference in India. Stephane was interested in what I was doing with FrontlineSMS, and was drawn to my focus on last mile communities, and the kinds of places where communication was a challenge. In what became my first real conference speech, I was invited to Bangalore that December where, in addition to talking about appropriate technologies, I chaired a meeting on networking technologies. Despite the distance – 20 hours stuck on a plane each way – the event was a great early experience, and Stephane and I would become good friends, later co-chairing the W3C Mobile Web in Developing Countries Group together.

Everything began steadily as we entered the new year, and I felt refreshed after a relaxing Christmas break with Elina, who had flown over from the UK. A road trip to Las Vegas, a place I swore I'd never go, turned out to be a surprising highlight. The number of visitors to the kiwanja.net website continued to grow after I added a database of mobile projects and a gallery of mobile phone photos and, come March, work began on the creation of The kiwanja Foundation, which would come in handy later as my work took off and began to attract funding. That month I gave talks at IDEO's head office in Palo Alto, at an event in San Francisco, and at a university in Tucson, Arizona with Edgardo, one of the other Fellows. I was also invited back to India, this time with Amnesty International, who were interested in the use of FrontlineSMS by local campaigners working to halt the unregulated sales of arms.

March had been particularly frantic, but it was nothing compared to what happened next.

Late one evening, during that trip to Arizona with Edgardo, I'd missed a call from Nigeria. I thought nothing more of it until an email arrived on 17th April from a group calling themselves the Network of Mobile Election Monitors (NMEM). They were preparing to use FrontlineSMS to monitor the Nigerian presidential elections taking place in less than a week's time. This would be the first ever large-scale use of FrontlineSMS that I knew of, anywhere. 'They must be mad', I remember thinking to myself. It took a while for their plans to sink in, but once I'd read their press release a couple of times and realised they were serious, I began thinking of ways I could help them, over and above providing technical assistance on the day. Crucially, for their monitoring efforts to succeed, they'd need people to text observations into one or two central mobile numbers. Those numbers would need to be promoted widely, something they'd already started doing themselves through local groups and other civil society organisations. I reached out to a few contacts I had in the country and did my best to help, but realised we'd need something bigger for the monitoring to really take off.

Could that 'something bigger' come courtesy of the BBC, I wondered? I was still in touch with Bill Thompson, whom I'd first met during that technology event we'd run in Cambridge a few years earlier. I dropped him an email, excitedly sharing news of what NMEM were planning, and he replied almost immediately. He promised to pass it on to a colleague who ran the technology pages of the BBC News website. Things were going to be tight, but at least they were moving. By now the election was only a couple of days away. I held my breath.

The election itself was a big deal and, because we're talking Nigeria here – one of the largest and most populated countries on the African continent – it was always going to be on an almost unimaginable scale. During a seven-hour window on Saturday 21st April, over 60 million registered voters would spread out across the country and cast their votes through one of 120,000 ballot boxes. Nigerian elections are always a major event, but this one was particularly significant given it signalled the first time since independence, almost 50 years earlier, that power would be handed over from one democratically elected leader to another. NMEM were hoping to reach half the electorate, who would be encouraged to text in their observations on election day. They hoped having ordinary citizens join already established international monitoring efforts might help give a more complete and comprehensive picture of how election day went, and highlight what worked and, crucially, what didn't.

With just two days to go I was contacted by Jon Fildes at the BBC. I'd already given Bill all the information I had, but Jon wanted to speak with me to fill in a few gaps, get a bit of the history of FrontlineSMS, and grab a couple of quotes. I also put him in touch with Emauwa, who was running the project for NMEM, and Jon was attempting to reach out to him for similar input, albeit with a little difficulty. NMEM were the ones doing the work, after all, and the story was theirs to tell, not mine.

Another day passed by. Challenging time differences with the UK, and communication problems in Nigeria, meant that things were now getting incredibly tight, with polls opening in less than 24 hours. By the end of that Friday I'd spoken to Jon and done all I could, so I signed off and headed to the Rose & Crown for an end-of-week beer with Erik and some of the other Fellows. I grabbed a spot of late dinner and made my way back to my camper to sleep, hopeful that the following day would be a good one for all my Nigerian friends. I'd not know for a few more hours whether or not the BBC had managed to get their story together in time.

I woke up early that Saturday morning, around 4:30am on the West Coast, and took the short walk from Stock Farm Road car park to Cordura Hall. In Nigeria the polls had already been open for a few hours, and would soon be closing. As I walked in the fresh, early morning air, the sky only just beginning to brighten, I wondered how things had gone for NMEM. And I wondered about the BBC. After election day none of this would be a story of interest any more. The window of opportunity was always going to be incredibly narrow. Had we made it?

I needn't have worried. 'Texts monitor Nigerian elections' was the headline, the top story on the BBC News Technology page. 'Anyone trying to rig or tamper with Saturday's presidential elections in Nigeria could be caught out by a team of volunteers armed with mobile phones.' I stared at the story for an age, slowly coming to terms with the significance of everything, not just for NMEM and the growing interest in citizen election monitoring, but for FrontlineSMS itself.

It had been a full 18 months since that rather haphazard launch, and I was proud of every single user we'd gained in that time, and proud of all the work they'd been doing with it. Not all of their stories made headlines on the BBC website, but they all mattered to me, nonetheless. I'd been aware from the very beginning that focusing on smaller, grassroots organisations was unlikely to lead to the sort of headlines that larger organisations would crave, but I was happy with that. Simple, appropriate technologies in the hands of those closest to the problems was all that interested me, and I now had a technology that lived and breathed that very approach. That weekend NMEM had just taken it up a level. Several levels, in fact.

I couldn't have been happier, or prouder.

Imperial

I had already been speaking to the MacArthur Foundation for a few months by the time the Nigerian election news broke. In fact Jerry, one of their programme officers, had reached out to me just a couple of weeks after FrontlineSMS first launched in 2005 saying he liked the work I was doing. I didn't realise the significance of his approach at the time – it's not every day one of the largest foundations in America emails you out of the blue like that, after all – but because I had zero knowledge of fundraising, and wasn't actually looking for money, I thanked him and casually tucked it to one side. I decided to reconnect with Jerry when I got to Stanford, and we were now chatting about possible funding to build a new and improved version of FrontlineSMS. Things had been moving a little slowly, but the news from Nigeria, and that BBC headline, did wonders to speed things up. There was still a fair amount of back and forth before I got the cheque, though. There was an expensive lunch at a fancy Stanford restaurant, followed by a low-key, early morning breakfast, this time in Palo Alto at a cafe made famous by the number of Silicon Valley startups that had sealed big funding rounds there. It was wonderful to be making my own sort of history, albeit on a much smaller scale, over fresh coffee and toasted bagels with Jerry that morning.

When I look back at the five years that followed the 2007 Nigerian elections, everything that happened feels more than a little surreal. During that time FrontlineSMS quickly drew users from every corner of the globe, positively impacting the lives of tens of millions of people. That, in turn, brought funding, and many writing and speaking opportunities, plus a huge amount of travel and an avalanche of awards, unimaginable just a short time earlier. For a while everything I touched seemed to turn to gold.

It reminds me of a story I read once about one of my all-time favourite bands. During a prolific spell of chart success in the late 1980s, British pop duo the *Pet Shop Boys* experienced what Neil Tennant, the singer, later described as their 'Imperial Phase'.

Over a three year period the band had an incredible run of form and, in Tennant's words, they could 'seemingly do no wrong'. For a while they were widely considered to be the biggest band in the UK. This is how it felt for me after the Nigerian elections. It was a time when me and my work could do no wrong, a spell during which FrontlineSMS went on the most incredible, brilliant run.

I didn't know it yet, but I was about to enter my very own 'Imperial Phase'.

The mid-2000s were an exciting time to be working in tech, and thanks to my location in the heart of Silicon Valley I was rarely far from the action. A short walk down the road from my Stanford office, at the top of University Avenue, sat the offices of an up-and-coming startup by the name of Facebook. I passed their modest looking glass-fronted office regularly as I made my way towards Peet's Coffee & Tea, one of my favourite remote working spots. Late in 2006 Facebook had decided to open up their social media platform after limiting access to high schools and universities for the first couple of years, and things were beginning to take off. At the end of 2006 they also added much-needed mobile support, giving their growth a further boost. Things at the company were still relatively low key and informal back then. Mark Zuckerberg, one of the founders, was a regular customer at the Indian cafe at the far end of University Avenue where many of us used to go in the evenings. I never saw him eat there myself, but a couple of friends did. I bet he doesn't go anywhere near places like that anymore, at least not without bodyguards.

It wasn't lost on me that, while I was focusing all my time, energy and effort on a free platform that was never likely to make any money (but that had the potential to do enormous social good), someone down the road was focusing all their time, energy and effort on something that would end up making them ridiculous amounts of money (and, ironically, end up causing enormous social harm). Financially, late 2006 would have been a smart time to join a company like Facebook, too, given how big it would become and how handsomely those early joiners would benefit when it floated a few years later. Funnily enough, I did attend a Facebook recruitment event on Stanford campus one afternoon, more out of curiosity than anything, but left with only a few keyrings and a bag of mints. Not even Facebook could tempt me away from FrontlineSMS back then.

That said, clearly money was never much of a priority given my choice to live in a van, and how I'd turned my back on promising careers over and over again in my earlier search for purpose. But I do sometimes wonder whether I should have focused on getting even just a little bit rich before embarking on a career in tech-for-good, money that may have helped me make a better, more sustainable go of it. The more I came into contact with tech philanthropists, many of whom made crazy amounts of money before developing a conscience, the more I realised how much more people can do with a steady stream of cash behind them.

At least I was now starting to get money to spend, though, even if it wasn't my own. On 29th May 2007, Jerry's confirmation email finally came through. 'I think it would be safe to say you have an approved grant from the MacArthur Foundation for $200,000'. Exactly 18 months after its rather low key launch, FrontlineSMS finally had some proper money, and a tidy sum at that, even after Stanford University had taken their not-insignificant cut. My year as a Collaboration Fellow was about to come to an end, and a bonus year as a Visiting Fellow was about to begin.

The breakthrough use of FrontlineSMS, and the MacArthur funding, were both entirely unexpected and I already had a flight booked home for the summer. Although my Stanford adventure would now go on a little longer, I still decided to head back for a break, or at least what I thought would be a break. A friend of a friend allowed me to park my van on his land up in the hills for a couple of months, and I returned to the comfort and familiarity of my cosy little flat in Cambridge where I would take stock and plan the year ahead.

I was hoping to do a couple of things with the MacArthur money. First was to develop an updated version of FrontlineSMS, one which would run not just on Windows but also on Apple computers and the Linux operating system. Build to these three and you have pretty much all bases covered, meaning almost anyone anywhere would be able to run the software, whatever kind of computer they had. To achieve this I decided to call in Masabi, the company that worked with me on the Silverback game four years earlier. We would use Java, the same language used for the game, which would allow us to write one single program that would run seamlessly on all three platforms.

The second was to build a new website. The one I'd quickly thrown together in a panic on that kitchen table in Finland didn't really cut it any more, and if we were going to take full advantage of this new opportunity we'd need something that communicated the offering a heck of a lot better. I'd already started approaching web companies in the Bay Area for quotes when I received an unexpected Facebook request one evening from a guy called Renny. 'Ken – came across kiwanja.net and am impressed as hell. GREAT STUFF. I head up digital strategies at Wieden and Kennedy, and we will be working with Nokia in developing countries.' I'd never heard of Wieden and Kennedy before, or W+K as they're better known, although after a little digging I was stunned to find out they were one of the biggest and most successful design agencies in the world, and the ones responsible for the iconic Nike 'Just Do It' tagline, no less. I accepted the friend request and Renny and I jumped on a call a couple of days later.

'Renny! So great to hear from you. Thanks for reaching out.'

'Hey Ken, I'm so excited to have found out about your work. I really love what you're doing with kiwanja and FrontlineSMS. We really need to find something to do together.'

Renny and I chatted for a while. It was a fun, easy-going, free-flowing conversation, at times resembling a mutual admiration society. We were both equally excited about stumbling across each other's work, but we still needed to find something we could do together. I did have one idea, but didn't hold out much hope that it was a good one.

'Well, our funding is generous but, given our ambitions, it's already a bit of a stretch', I explained. 'I am looking at a new website and have been speaking to a few agencies, but don't have anything concrete yet. Maybe we could look at doing that together.'

'What's your budget, Ken?'

'I think we're looking at around $30,000 max, Renny.'

Given the first FrontlineSMS website cost me nothing, and I'd built the kiwanja.net website myself, too, I wasn't used to spending any kind of money on this sort of thing. But I knew we'd need expert help this time around and, despite being confident in my own abilities, I'll be the first to admit that this was one task best left to the professionals. W+K definitely fell into that category.

'Leave it with me, Ken. I'm sure we can figure something out.'

'Thanks, Renny. I'll keep my fingers crossed.'

I took full advantage of the little time I had in the UK, making regular visits to the Masabi offices in London to help out with the FrontlineSMS rewrite. By now they'd hired a new developer, Alex, who was leading the work for them. Within a couple of years we'd hire him ourselves, and he'd end up working firstly out of our London, and then our Nairobi, offices. FrontlineSMS was already proving a huge draw, attracting all manner of incredible people to the cause. Hiring talent had never been easier.

As my work grew so did my visibility, and I found myself drawn more and more into other people's work. Never one to turn down an opportunity, I grabbed at every single one. Summer 2007 was already crazy enough but, to add to the chaos, I agreed to work with Grameen Technology Centre on a new initiative they were putting together in Uganda. In collaboration with Google, the AppLab project was looking to explore different ways mobile phones could be used to share information with communities, anything from health advice to the price of goods at local markets. SMS was still the primary delivery method for most people back then, so the work was a perfect fit for me. Grameen and Google certainly seemed to think so.

Before joining them in Uganda there was the minor matter of Elina's graduation, and then a wedding, to deal with. A few months earlier, Elina and I had got engaged and we'd set a date for the summer. Clearly, nothing was going to stand in the way of that, and come July we spent a wonderful day in Brighton with friends and family from Jersey, the UK and Finland, getting married in the grand surroundings of the Royal Pavilion building. The whole day was kept low key and simple, something we both wanted. It was enough to be back in Brighton, the place where we had met and studied, and to be with friends and family. That Christmas we planned for Elina to fly over and visit me in San Francisco, and we'd take the chance to grab a late honeymoon in Hawaii. For now, she was busy preparing for a new job, and I had that trip with Grameen and then a return to Stanford. As it turned out, a delayed honeymoon suited us both perfectly.

Clearly the timing of the Grameen trip wasn't great, but I was grateful for the opportunity to get back into the field. It was the place I was the happiest, after all, surrounded by the very communities I always hoped to help. David, the director of the AppLab project at Grameen, would be with me for the first week, after which he'd head back to Seattle, leaving me to continue with the research, largely alone, for the next three. David was a tall, smart and articulate guy, with a long and successful career in tech, and our shared sense of humour and determination to enjoy everything we did meant we had a huge amount of fun working together. We still joke about me spending my honeymoon with him, which, in reality, wasn't a million miles from the truth. Other colleagues on that trip worked for MTN, one of the larger mobile operators across Africa, and Google, who were keen to understand the opportunity in emerging markets. Uganda was to be something of a test bed for them.

I flew into Kampala from London in early August, just a few days after getting married, and met David at the guesthouse where we'd both be staying for that first week. I walked into the dimly lit garden restaurant and was immediately at ease, thankful for the fresh breeze and grateful for the calmness, and the abundance of green vegetation growing effortlessly up the walls. I waited for a wave or a nod, or for someone to stand up and look my way. David and I had only spoken on the phone for a short while, and this was the first time we would actually meet. I wasn't sure who I was looking for.

'Over here, Ken!' came a shout from the far corner. I smiled and walked over. Before I could reach out to shake David's hand, the waiter jumped in and asked if I wanted something to drink.

'Oh, a Nile beer, please.' I decided a celebration was in order, and it had been a long day of travel. And it was hot. And I should technically have been on honeymoon. David and I had got to know each other fairly well, as much as we could on a one-hour phone call anyway, but once we actually met our friendship would quickly blossom and continue for years after the trip.

'Great to see you, Ken. Grab a seat.'

'Thanks. Really good to finally meet. It's great to be back in Uganda, too.'

My Nile quickly arrived, balanced precariously on a chipped plastic tray complete with a freezing cold, frosted glass with the name of a completely different brand of beer on the side. But I didn't care. The waiter poured about a third and then stopped, a strange habit, I always thought, shared by waiters the world over. I poured the rest, and David and I raised our glasses. After a hectic few days and a long flight, a cold beer was just what the doctor ordered.

'Here's to spending your honeymoon with me', joked David.
'Indeed', I laughed back. 'Aren't you the lucky one?!'

I've always believed the best time to get involved in any kind of technology-for-development project is right at the very beginning, that briefest of times when everything is on the table, nothing is ruled out and there's no such thing as a bad idea. I often describe the earliest days of the sector as a 'golden age of discovery'. It didn't matter what you were doing, there was a good chance it hadn't been tried before. Almost everything was new, exciting and groundbreaking, and for those who were committed and brave enough, blazing a trail had never been easier.

Technology aside, I also enjoyed the process of settling down in a country, getting as close to the action as possible, and spending as much time as I could in the company of local people. Maybe it's just the anthropologist in me, or the fact I've always been curious, inquisitive and something of a people person. Who knows. But I've learnt that, if you take the time to look, there are rich sources of information everywhere, everything from the usual TV, newspapers and radio to conversations with taxi drivers (who, regardless of where they drive seem to have answers to all the world's problems), villagers, market traders, waiters and children – even eavesdropping conversations in bars – all of which can help build a valuable picture of what matters to people and, just as importantly, what doesn't.

While most of my Google colleagues were escorted around by ex-military wherever they went, staying in the best hotels in Kampala and eating in the fanciest expat restaurants, I was happiest out roaming alone, exploring Kampala on foot as I made my own private map of the best local places to eat, the best book shops, and the best places to buy local antiques and arts. I preferred to stay in more modest surroundings, the sorts of places where you could get a real sense of what life was like for ordinary people. I felt sorry for some of my colleagues who, due to company policy, weren't allowed to do anything like this.

For the first week of this particular trip I'd be staying in the same place as David, a pleasant little guesthouse on the hills overlooking Kampala. But once he left I lost no time checking out in search of somewhere a little closer to the action. I spent that first Saturday pounding uneven pavements, checking out hotels and guesthouses close to the centre of the city. My accommodation budget for the next three weeks wasn't great, but neither were many of the rooms I was being enthusiastically shown by the owners. Some seemed to think the presence of a bedroom door was a selling point. Other rooms had no glass in the windows, or broken taps in the bathrooms. Each viewing was followed by an almost immediate, yet polite, rejection. I wasn't expecting the Ritz, but I did want somewhere safe, secure and reasonably comfortable to stay.

A few failed attempts later I settled on a room in the Tourist Hotel, a scruffy but modern-looking building on a busy junction overlooking Nakasero Market. Crucially, at just under $50 per night, it was in budget. The rooms were basic but at least they were clean, the hotel was safe and secure, and the rooms had doors. There was a large, open-plan lounge area on the ground floor, with a good-sized television that regularly showed live Premiership football, and I'd enjoy watching a few games there with some of the other guests, most turning out to be Chelsea fans, oddly enough. There's nothing quite like football to bring people of different cultures and backgrounds together.

Fortunately for me, my single room was near the back of the hotel. The front faced straight onto Nakasero Market, a cultural hotspot that never slept. The all-day and all-night clubs and bars, the constant humming and buzzing of traffic and the animated shouting and screaming of passers by, were all things that most sane people would rather avoid having right outside their bedroom window. During the day cars, bikes and lorries squeezed into every inch of road, filling the air with heavy, suffocating fumes. The market itself was an explosion of sound and colour, with almost everything you could imagine on sale from fruit and vegetables to fresh meat, clothes, cloth, carpets, mobile phones, cheap Chinese radios, wristwatches and dubious-looking bootleg DVDs. A selection of chicken, beef, corn and cassava sizzled on charcoal grills dotted between the corrugated roof sections, plumes of light grey smoke lifting on the gentle breeze, carrying the smell towards unsuspecting customers on the streets and dirt pavements outside. The constant chatter of daily life, of deals being done or arguments being settled, filled the air, occasionally interrupted by the heated exchange of animated customers bartering for goods. This was Nakasero Market, and there was never a dull moment. It was exactly what I was looking for.

Over the course of the month I managed to come up with a few dozen ideas for potential mobile services, each based on my growing understanding of Ugandan life. I'd present these to Google and Grameen in Seattle later in the year, during a brief break from Stanford, and they'd work on what they believed to be the most promising ones. AppLab finally launched five mobile services in June 2009, almost two years after my work had ended. Despite having the smartest people, generous funding and the best technology resources available anywhere, the time it took to get some of these ideas to market sat in stark contrast to how nimble and agile things had been for me with FrontlineSMS. I enjoyed both approaches, but if given the choice would always prefer getting things done quickly, allowing real-world users to shape the direction the work would go. I'd be invited back to Uganda one final time, a few months later in December, to provide technical assistance as Google and AppLab began testing out their early ideas and prototypes.

Summer 2007 had been non-stop, and as autumn approached things weren't about to slow down. Days after leaving Uganda I managed to squeeze in another trip, this time to Nairobi, to speak at a 'Mobile Activism in Africa' event. A number of news outlets had jumped onto the Nigerian election monitoring story, too, and I'd end up speaking to *The Guardian*, the BBC World Service and *Falmer*, the Sussex University magazine, among others. Another prominent Kenyan blogger, Erik Hersman, interviewed me for his 'African Digerati' series, an early conversation that would lead to a lifelong friendship. FrontlineSMS was shortlisted for a prestigious Mobile Messaging Award, and kiwanja.net was appointed a Stockholm Challenge Champion. Interviews with the Corporate Council of Africa followed, as did others for Nokia and *The Economist*. Further trips to a roundtable event in Aspen, Colorado, and an Internews event in Istanbul, topped off an intense couple of months. Things may have been busy, but after years in the wilderness I wouldn't have had it any other way.

While I was ducking in and out of conference rooms, FrontlineSMS continued to make itself useful, helping monitor another national election, this time in the Philippines. Activists in Pakistan also deployed it to help get information safely out of the country during General Musharraf's controversial six-week State of Emergency declaration in early November. In the middle of it all, I managed to launch a competition for grassroots nonprofits called nGOmobile, offering free laptops (donated by Hewlett Packard), GSM modems, and copies of FrontlineSMS, to organisations that identified a compelling use-case for SMS in their work. I seemed to relish the craziness of it all, constantly coming up with ideas and acting on them with little regard for how few hours might be left in the day. Being busy became a habit that was hard to let go.

Self-inflicted distractions aside, FrontlineSMS remained my main area of focus and, thankfully, Masabi had been plugging away diligently in the background. We'd remained in close contact over the summer and winter, and I'd get the odd 'Alpha' release which I'd test and provide feedback. By the end of the year Renny had also come good with his promise to get involved, and as 2007 closed out we announced the awarding of the FrontlineSMS website contract to Wieden and Kennedy. 'W+K believes in the power of mobility, and the FrontlineSMS mission, and we're very proud to be part of this effort', said Renny the day we shared the news.

We were aiming for a big launch in the summer, ideally June or July 2008, around the time the MacArthur money would run out. We'd need to have the new version of FrontlineSMS written and fully tested by then, and the new website designed, developed and live. The six months leading up to the big day would see everyone get through a huge amount of work, and I was grateful to be supported by two companies so completely committed, and professional, in everything they did. FrontlineSMS was already beginning to take on a life of its own, possessing an uncanny ability to motivate, inspire and draw in almost anyone who came into contact with it. I just sat in the middle, like a conductor, coordinating everything as best I could, while Masabi and W+K weaved their creative magic.

Before the new revamped FrontlineSMS and FrontlineSMS website would go live, conference invites continued to flood in. All manner of people and companies, it seemed, were keen to hear the story of this scrappy bit of software. Nokia and Vodafone wanted a piece, as did Sussex University, *The New York Times* and Canadian Radio. Offers came in to speak at Mobile World Congress in Barcelona, and at the Emerging Communications Conference in Silicon Valley. A professor at Stanford, BJ Fogg, asked me to share my work at his Texting4Health Conference, and Microsoft Research in Cambridge dragged me all the way home to give a presentation at an 'intimate staff gathering' they were organising. Further keynote addresses at the Global Messaging Conference in Cannes, and at a W3C Mobile Web event in Brazil, helped take the buzz and excitement ahead of our summer launch to dizzying new heights. Looking back, it's exhausting just thinking about all the promotion opportunities we took advantage of back then, and we hadn't even launched our shiny new product yet.

Renny being Renny, and W+K being W+K, a decision was made that we needed a new logo for FrontlineSMS. The small mobile phone icon I'd come up with at the very start, the one still glued to the Million Dollar Homepage, simply didn't cut it anymore. It would have looked embarrassing on the beautifully crafted website they were intelligently and sensitively putting together for us. It was something they wanted to put right.

'How do you feel about some new branding for FrontlineSMS?' asked Renny during a call we'd arranged late one afternoon.

'Sounds amazing, Renny! What did you have in mind?'

'Well, budgets and time are a little tight, but we've chatted here and think we can come up with a fantastic new logo for you. We've got a few questions to help us frame our thinking. Do you want me to send them over?'

Never in my wildest dreams did I think I'd ever work with some of the most talented brand experts in the world, the same people more used to partnering with corporate giants like Nokia, Samsung and Nike. Unsurprisingly it was a fascinating and eye-opening process to go through, something far more structured, and with far more depth, than anything I'd ever experienced before. An early objective was to come up with a list of keywords which we felt best described the software. One immediately stood out, one which not only happened to be central to my earlier thinking, but which also came through time after time in messages from the growing community of users. That word?

Empowerment.

It was perfect. Empowerment is hugely personal and emotive, and luckily for us it's also something that can be expressed physically. How to graphically represent this 'physical expression of empowerment' became a key theme as logo ideas began to emerge. The neat concept of a 'textable logo' was also beginning to take shape, something that would later prove to be something of a masterstroke.

The W+K team were loving the work, and seemed to thrive focussing on something a little different, something they felt might make an actual difference in the world. Renny continued to speak fondly of the opportunity to work on something that felt truly transformational. 'Ken built FrontlineSMS out of love, faith in human potential, and an inspired application of mobile technology', he said at the time. 'And you can feel it when you talk and work with him. At W+K, while we have the privilege to work day in and day out on some pretty impressive brands, the chance to help craft the visual language and web experience for Ken's creation was uplifting. From our first conversation with Ken, W+K has felt like a part of the extended FrontlineSMS family.'

By June, over 25 organisations had volunteered to help put the new version of FrontlineSMS through its paces, giving us one final chance to iron out any last-minute bugs before the big launch. W+K had finished the website, which looked beautiful. What's more, the logo they produced – \o/ – was the perfect representation of empowerment, a textable 'image' of someone with arms raised in the air. Over the coming years we'd get a steady stream of users sending in photos of their teams, or the communities they were supporting, with their arms outstretched above their heads. I'm sure most academics and development professionals would just shrug this off, but for me this simple act signalled a wonderful, personal, priceless connection that we'd made with these people, people desperate to be given a chance to try out their own ideas, and solutions, to put an end to the problems many of them were facing in their communities. To this day, being able to support their work remains, for me, one of the greatest privileges of the whole FrontlineSMS era.

The big day finally arrived. Wednesday 25th June, 2008. A year in the making, we were finally ready to unleash a revamped FrontlineSMS, a shiny new website and our amazing new branding on an unsuspecting world. Compared to the chaos of the previous few months things were surprisingly calm and sane that day, and the news was steadily picked up by many of the technology and activism sites around the web. It felt like something of a new beginning, a bit of a rebirth, and downloads and use-cases began picking up as curious organisations grabbed their own copy to see what the fuss was all about.

The day of the launch was also a great excuse to break news of another exciting deployment of FrontlineSMS, this time supporting community healthcare workers in Malawi. Just over a year earlier, a few days before the Nigerian elections in fact, I'd been introduced to Josh Nesbit, a Stanford human biology undergrad, and we met for lunch one afternoon in the shade of one of the imposing Californian oaks dotted around campus. I asked what he was doing, and what I could do to help.

'I've been awarded this Fellowship, Ken, and next summer I'll be returning to St Gabriel's Hospital in Malawi to help out.'

'Oh, so you've been there before?'

'I have. Our family has a long-standing connection with the hospital, and I'm looking forward to going back. I've heard about your work, and have been wondering whether your SMS system might be something they could use.'

'I'd be happy to help you find out. If you're looking to solve any kind of communication challenge, FrontlineSMS might be a great fit.'

St Gabriel's Hospital was woefully understaffed, with just two doctors and a handful of clinicians serving a whopping 250,000 people over a 100-square-mile catchment area. Given the lack of resources, the hundred-odd community healthcare workers (CHWs), mostly volunteers, were a vital component in the delivery of any kind of care. Once they left the hospital gates, though, they had little to no communication with the doctors until they returned, usually around three months later. Josh realised that if he spent his month in Malawi setting up FrontlineSMS at the hospital, and armed the CHWs with recycled, basic mobile phones, he could create a powerful communications network that might solve the problem.

His own efforts aside, Josh turned out to be a fantastic ambassador for FrontlineSMS. Not only was his work in Malawi a resounding success, it was inspirational, and a story people couldn't seem to get enough of. Even his father, a prolific author with well over a dozen books to his name, ended up weaving an SMS communications platform into the storyline of his 2010 book, *Peace*, and made a particular mention of his 'relentless, positive storm' family in the acknowledgements.

Traditional news organisations were also all over his work, including CNN who produced a short film. Josh immediately saw the potential in what he was doing, and very quickly started assembling a team of his own, making a return trip to St Gabriel's in late 2008 to share new ideas and check how things were going. The response, feedback, level of excitement and impact were all more than he could have wished for. By this time he'd also been joined by Isaac, another student, who would come aboard as one of the co-founders.

Josh and Isaac quickly saw the opportunity to build on this early work. Their first move was to develop a new module called 'Patient View' to run within FrontlineSMS. 'Patient View' would allow patient records to be shared, and updated, remotely by text message. Alex, who by this time had left Masabi and was working for us full-time, provided excellent support as they designed, developed and tested the new functionality. Everyone was happy, positive and motivated, including Josh, and in early 2009 he approached me out of the blue with an exciting proposal.

Not realising, or perhaps expecting, his idea to take off the way it did, Josh had settled on 'Mobiles in Malawi' as the project name. As interest picked up and his young team started to grow, he decided they needed a name that was a little more catchy, something that better leveraged the FrontlineSMS brand and reflected what they were doing with it. Their suggestion? FrontlineSMS:Medic. They weren't sure how I'd feel about sharing my brand, but I loved it. Little did they know, but that day they also started a trend that would see the FrontlineSMS family grow in ways none of us ever expected.

A few months later I received an email from Ben Lyon. Ben had met Josh a few weeks earlier and they'd quite literally sketched out an idea on the back of a napkin. 'I've been following your work for a while now and am excited to run some ideas by you. I'm currently working on a model for FrontlineSMS:Credit, which will use the FrontlineSMS platform to provide over-the-air microcredit to entrepreneurs in hostile and inhospitable areas.' Ben was keen to move forward, but wanted to be sure I was happy with his choice of name. By now FrontlineSMS seemed to be almost everywhere, and if I'm honest I was thrilled to support Ben's work, as I had been with Josh. Not for the first time, I gave the green light.

But that wasn't the end of it. We'd later see the emergence of FrontlineSMS:Learn, supporting remote learning via text message, and FrontlineSMS:Radio with its focus on transforming rural radio programming into a real-time, two-way conversation. Adding to the sense that all of this was meant to be, the W+K logo lent itself brilliantly to each of these new family members. \+/ for Medic. \$/ for Credit. \?/ for Learn. And \~/ for Radio. Completely by accident we'd become something of an incubator for a whole bunch of other talented young innovators and their ideas. It was a role I was more than happy for us to play.

FrontlineSMS:Medic would later become Medic Mobile, an organisation that today supports tens of thousands of community healthworkers in dozens of countries around the world. FrontlineSMS:Credit would eventually become Kopo Kopo, a venture-funded company that manages mobile money payments for thousands of East African merchants. FrontlineSMS:Radio would spur the creation of a new project at Cambridge University, Africa's Voices, that works with rural radio stations to help citizens in African countries participate in government decision-making. Only FrontlineSMS:Learn from those early days failed to turn into a fully funded, independent project or organisation of its own. All-in-all, it wasn't a bad record for something that happened completely by chance.

With so many people now joining me in sharing FrontlineSMS with the world, opportunities to write and speak came in with increasing frequency. There were requests for interviews, and invitations to write guest posts, or articles for websites, or chapters for books. There were conferences wanting me to fly over and share our story with their audiences. And there were incredible new use-cases that would elevate our work to yet another level, or into a completely new sector. Anything, and everything, seemed possible. At times it felt like we were blessed with the most perfect of projects.

One thing that blew me away the most in those early 'FrontlineSMS family' days was how, as a group, we all managed to lift one another. Good news for the core product was good news for the growing number of family projects. Good news for the growing family of projects was good news for the core product. And users who took these tools and did something brilliant with them not only elevated their own work by sharing it with us and the world, but they inspired others to try things out for themselves. People were proud, excited and motivated to apply FrontlineSMS in ways none of us had ever imagined, and the online community of users, which would number thousands at its peak, became – without a shadow of a doubt – our biggest achievement. To wake in the morning to find a user in Guatemala had helped a user in Pakistan gave me goosebumps. People seemed to go out of their way, not only to support one another but also to share their work, uploading photos and videos to the growing community site. I'd never seen anything like it in any mobile-for-development project before, and I don't think there's been anything like it since. Building genuine, engaged, empowered, motivated communities is hard, but somehow we managed to do it, and almost without trying.

For a while everything we did try seemed to come off, but storm clouds were gathering. Despite all the growth, I'd managed to keep project management low key and lean, running the relaunch between my Stanford desk, Peet's Coffee and my beloved camper. I still didn't have any overheads to speak of, except a modest, basic salary for myself. I'd avoided building any organisational structure around the platform, partly because keeping it lean felt like the right thing to do, and partly because I didn't want to lose the free-flowing, organic, opportunistic nature of it all.

All that aside, I did now have a problem, if that's the right word to use. Because I hadn't written the new version of FrontlineSMS myself, I couldn't personally provide the level of support I was used to, and couldn't fix bugs and release new versions alone, something else I'd been able to do before. I was now dependent on Masabi, and that always sat at the back of my mind. However engaged they may have been, they'd need paying to do any future work, and the MacArthur money was just about spent. And so was my time at Stanford. Not only was it time to start thinking about selling my beloved van, I had to start thinking about what might be next. Find money to maintain growth, or cut back?

Little did I know but a short walk across campus, sitting at his desk inside the towering Hoover Building, was Larry Diamond, a highly respected political scientist and an expert in democracy studies. Larry penned one of the first books on the failure of America's 2003 post-invasion plans for Iraq – or, more to the point, the lack of any. Larry was something of an expert on Nigeria, and it was the use of FrontlineSMS during their 2007 elections that first drew his attention to my work. A couple of months later, in early July, we were introduced by Kathy, a mutual friend in the Stanford anthropology department. (Kathy was the one who also introduced me to Josh.) Larry was fascinated by FrontlineSMS and, more broadly, how emerging technology might help strengthen democracy and democratic institutions around the world. This was to be the foundation of a new initiative he was working on called 'Liberation Technologies'.

Larry was a tall, rather imposing figure, gently spoken and thoughtful, smartly dressed though rarely in a suit. I was thrilled that someone of his stature was remotely interested in my work, and he was thrilled to have found someone working on the very technologies he was most interested in. Over the next few months we'd meet on campus regularly, at least as often as his hectic schedule would allow, and we'd occasionally go out for dinner in Palo Alto. On one occasion we both found ourselves in Washington DC at separate events, and we met up and went out for a curry. Larry became a good friend and something of an ambassador for FrontlineSMS, and I remain grateful to this day for everything he did for me in those early days. If FrontlineSMS ever needed help, I could always rely on Larry to step up.

And step up he did. By March 2008 the end of the MacArthur money was in sight. I had no idea what I'd do when it ran out, and I had no idea where any further funding might come from. I was new to all of this, after all. Larry was also worried that a project he considered 'one of the most exciting initiatives in the fledgling liberation technology movement' might disappear out of sight, and took it upon himself to reach out to a friend at the Hewlett Foundation, which was situated at the top of Sand Hill Road, just outside campus. Larry spoke highly of my work, and the importance that it continued, and his endorsement was massive. Hewlett liked what I was doing and they clearly trusted Larry's judgement. They were also impressed with the impact of the project so far given that funding, to that point, had been modest, and the guy running it all was running it alone, with few resources, and living in a van.

Conversations continued into the summer – these things always seem to take forever – but the signs were good that there might be an appetite to support my work. I didn't have a Plan B, so everything hinged on Hewlett. I arranged to meet up with a couple of their senior directors on my return to London, and hopefully there we could seal the deal. The final act of my extended American adventure would be to get rid of my van, signalling the end of one of the most productive and enjoyable periods of my professional life.

The van turned out to be an easy sell. I think it only saw rain once or twice during my entire stay, and it looked as good as the day I bought it from Rick. True to form, I put it on Craigslist and started getting interest straight away. After just two viewings I got the full $6,500 asking price. I'd managed to live in one of the most expensive parts of the United States for almost two years and was down less than $1,000 by the time it ended. Saying goodbye to the camper was hard, though, and sometimes I wished I'd found some way of keeping it, maybe shipping it back to the UK with me. At least I have photos, one of my favourites taken the day before I would finally let it go.

I soon arrived back in the UK, back home to my flat, but I didn't stay there long. Conference invites continued to come in thick and fast, everywhere from the World Economic Forum in New York to 'A Better World by Design' in Rhode Island. Sandwiched in between was a keynote address at an open source technology event in Berlin, a talk at the School of Oriental and African Studies in London, and a panel discussion at Social Capital Markets, all the way back in San Francisco. The Clinton Global Initiative (CGi) came calling, inviting me to their glitzy global summit in New York where I'd famously mess up my first photo opportunity with Bill Clinton. I had my eyes closed on that occasion, but did get a second chance a couple of years later and, much to President Clinton's amusement, did much better. CGi was the kind of event that made me feel more out of place than usual, the kind dominated by semi-famous high-net-worth individuals boasting about all the amazing things their family foundations were doing. Interview requests flooded in, too – with SHIFT Radio, Nokia Conversations, Radio France International, Danish Radio and *The Economist*, among a whole bunch of others.

Watching from a distance, as all this madness unfolded, were Andrew Zolli and Leetha Filderman. Andrew and Leetha were in the middle of planning the autumn launch of the first Pop!Tech Social Innovation Fellows Programme, and I had no idea they wanted me in it until, one afternoon during a brief gap in travel, the phone rang.

'Ken, it's Andrew and Leetha at Pop!Tech. How are you doing?'

'Oh, hi. I'm good, thanks. Enjoying the break in travel, if I'm honest! It's nice to hear from you.'

'Well, we have news of more travel if you're up for it.'

'One more trip probably won't hurt', I joked.

Pop!Tech were in the process of selecting 16 people to make up their first cohort of Fellows, and they wanted me to be one of them. There had been an earlier public call for applicants, which I'd missed, so their call really was out of the blue. I'd never heard of Pop!Tech before, but they'd heard of FrontlineSMS and they wanted me there to support my work, and to have me share it with the Pop!Tech community. The Fellows would convene just outside Camden, a picturesque town in Maine, a week before the annual Pop!Tech Conference, and then stay for the main event which ran over the best part of three days. I was thrilled. Stuff the travel, I thought. The conference alone sounded amazing. I said yes, of course, before Andrew and Leetha hung up, no doubt ready to ring the next unsuspecting Fellow on their list.

The week-long Fellows Programme was held at Point Lookout, a network of beautiful log cabins set in almost 400 acres of pristine forest on a hill overlooking Camden, with sweeping views of Penobscot Bay. Point Lookout was originally developed as a corporate retreat in the late 1990s by the credit card company, MBNA, and it had all the trappings of an executive playground – basketball courts, tennis courts, a bowling alley, restaurants and some of the most beautiful outbuildings and luxury cabins you'll ever see. As a retreat it didn't disappoint. It was an escape from the world at the bottom of the hill, a place to think, recharge, plan, plot, and forge collaborations and friendships away from the hustle, bustle and distractions of daily startup life.

On arrival the Fellows were split into pairs to share cabins. By pure luck my cabin partner would be Erik Hersman, a talented and respected Kenyan blogger whose work I'd learnt about early in my humanitarian technology career. We had a lot of interests in common and I, for one, was happy our names had been pulled out of the hat together. Our friendship would blossom over those two weeks, and we'd build a strong personal and professional friendship, one which continues to this day.

Erik had already built a solid reputation for himself long before my own work began to take off, but a national crisis in late 2007 would elevate him, his work, and his profile yet further. On 27th December that year, a contested Kenyan presidential election descended into chaos after Mwai Kibaki declared himself the winner. His opponent, Raila Odinga, cried foul, as did many international observers. Three days later, Kibaki was quickly and rather sneakily sworn in late one night, leading to numerous peaceful but noisy protests around the country. Some demonstrations turned violent, though, and police ended up shooting hundreds of protesters, some right in front of the watching media. This led to even more anger, bringing yet more people out onto the streets, and the country quickly spiralled into a cycle of violence.

Watching from his home in Florida was Erik. Watching from other parts of the world, including Kenya itself, were some of Erik's long-time African friends and colleagues. As the horrors unfolded right in front of their eyes and screens, they felt compelled to do something about it. In just a couple of days they hastily built a website (and later, an organisation) called Ushahidi, a Swahili word meaning 'witness', which they hoped would help capture details of what was happening around the country. People would report acts of violence, or acts of peace and reconciliation, and Ushahidi would plot the details on a map. Erik and his colleagues believed that effectively taking a picture of what was happening (and where) was the only way people might find out if their friends or families were safe, and the only way the country might reconcile itself once the violence was finally over.

Ushahidi was aware that, out of speed and convenience, most people would submit their reports by text, and he approached me about integrating FrontlineSMS into the new platform to help collect the messages. Within no time at all the site was up and running, and international interest grew almost immediately. Ushahidi was one of the first examples, if not *the* first example, of the potential of crowdsourcing humanitarian data. They would later have what you might safely describe as their breakthrough moment. In the aftermath of the devastating earthquake that hit Haiti late in the afternoon on 12th January, 2010, an Ushahidi map helped coordinate much of the global humanitarian response. FrontlineSMS proudly played a support role then, too.

Pop!Tech had been running for 22 years by the time Erik and I got our invites, and by then it had become one of the longest-running and most highly respected events of its kind. The brainchild of John Sculley, former CEO of Apple, and Bob Metcalfe, inventor of Ethernet (the technology used to connect computers together across a network), the conference began life as an excuse to invite their friends to Camden, Maine, where they both had homes. October was a particularly beautiful time in New England, with the onset of autumn and the explosion of brown, red, orange and yellow, and they wanted as many people there as possible to enjoy it with them.

As the event grew, Pop!Tech somehow managed to retain the feel of an intimate, friendly gathering, with speakers and guests staying close to one another in small local B&B's and eating and drinking in many of the same local restaurants and bars. Everything was within walking distance, with the local opera house at the centre, a stunning venue that gave the main three-day event a sense of ceremony and grandeur that's hard, if not impossible, to match in any modern hotel in any modern setting. High profile speakers would often join delegates at intimate parties and nightly gatherings, something that helped create an unrivalled, tightly knit community of thinkers and doers who kept coming back, year after year. It's no overstatement to say that Pop!Tech was one of the most brilliant conferences I've ever been to, and that first year I had a front row seat as one of a few lucky Fellows. Later, Ben and Josh would receive their own Fellowships on their own merits, and they would stand on that same stage in front of several hundred guests, sharing their own stories of how they'd used FrontlineSMS to help kickstart their own life-changing ideas and careers.

Pop!Tech came and went far too quickly for my liking, but I felt physically and spiritually refreshed by the whole experience. No sooner had I returned home when news came through that the Hewlett Foundation had awarded $400,000 to support the continued growth of FrontlineSMS. This time the money would last two years and it would be the first funding to go through The kiwanja Foundation, the organisation I'd set up during my time at Stanford to help raise funds for the project. Crucially, there would be no hefty cut for Stanford this time, either. Finally I could plan ahead for more than just a few months.

Thanks to all the publicity, FrontlineSMS downloads continued their upward trend. In the DRC it was used in conjunction with a popular TV series to help tackle the spread of HIV/AIDS, and in Malawi the media deployed the platform to help track their national elections. It supported the monitoring of the 2008 Zimbabwean elections and, in East Africa, it was used with Ushahidi to document places where essential, basic medicines were out of stock. To top it all, Brad Pitt and Angelina Jolie's foundation funded a FrontlineSMS-powered project in Cambodia, supporting vulnerable young children. It seemed that everyone wanted a piece of the action, including Hollywood film stars and, as it turned out, politicians, too.

Among the many surreal moments from those early FrontlineSMS years, the one where I found myself in 10 Downing Street having coffee with David Cameron, the UK Prime Minister, has to be right up there. I'd been invited to a meeting to discuss a forthcoming government trade mission to Africa, a five-day whirlwind trip to South Africa, Nigeria, Rwanda and South Sudan (which, thanks to an earlier referendum, was about to gain independence from the Republic of Sudan). What my mum would have thought of it all I'll never know. I was desperate to share it all with her, but she'd sadly passed away a few months earlier.

The famous black, shiny door opened slowly and I remember hesitantly stepping inside. I introduced myself to a member of staff and was issued with a badge. 'Ken Banks, Founder of kiwanja.net and FrontlineSMS.' Another member of staff led me down the carpeted hallway and up a flight of narrow stairs at the back. Paintings of prime ministers going back centuries adorned the walls and added to the incredible sense of history. We were ushered into a meeting room that was, by the time I got there, full of smartly dressed people chatting quietly, drinking tea and coffee.

Shaking hands with attendees as he made his way around the room was Prime Minister David Cameron. He was dressed smartly in his usual dark blue suit, and I stood there for a moment, trying to take it all in. Not bad for a lad from a council estate in Jersey, I remember thinking to myself.

Before I knew it, it was my turn.

'Thank you for coming', he said, extending his hand.
I reached out and shook it. 'Thank you for inviting me.'
He peered at my name badge. 'Ah, you're the person who developed that text messaging technology.'
'I am', I replied, briefly explaining how it worked and where it was being used, making sure I made a particular reference to its use in Africa.
'Mobile technology is so important in these places, isn't it?'
'Absolutely', I replied, nodding.
'It's got to be the future', said the Prime Minister. 'And it's key for Africa's development. That's why we thought it might be a good idea for you to join the delegation. We also thought it would be good to have someone from the nonprofit world. Most of the other delegates here represent big business.'

The delegation was meant to travel for five days, but five days is a long time in politics and we never did make it past our second scheduled stop. After 24 hours in South Africa, and another day in Nigeria, our chartered Virgin Airlines plane was on the way home, a major phone hacking scandal forcing David Cameron to return early. I remember little about the flight, but did enjoy air traffic control giving us priority as we approached Heathrow, our plane jumping the queue ahead of dozens of other aircraft circling that morning. Having a Prime Minister aboard does that, I guess.

Conference and workshop invites showed no sign of letting up once I was back. I spoke at an IREX event in Washington DC (where I shared the stage with Josh), did a demo of FrontlineSMS and Ushahidi in Doha (where I shared the stage with Erik), gave a keynote address at a mobile web event in Mozambique, spoke at a Global Engagement Summit at Northwestern University in the US, gave a talk at the 'Human Rights Centre' at UC Berkeley and another at a 'Computing at the Margins' event at Georgia Tech, took part in a panel discussion at a Vodafone event in London and another at 'Thinking Digital' in Newcastle, delivered a keynote speech at the launch of the UN Humanitarian Technology Challenge event in Washington DC, and shared my experiences at a UK Parliament committee gathering. There were also more interviews with Public Radio International, *The Independent* newspaper, the BBC World Service, Fast Company, *The Economist* and *The Guardian*, to name just a few. And there were guest posts for various sites and publications, including a chapter contribution for a Boston Review book on technology in Africa, and a guest article for *WIRED* magazine. This is how things would roll for the next couple of years, a relentless stream of talks, international travel, media interviews, writing, not forgetting user support and essential project management duties. It all felt absolutely worth it, though, and every day I woke up feeling incredibly grateful for everything that was happening, and for all the interest and support I was getting for my work, and for all the people it was helping. I never stopped reminding myself how lucky I was, and how few people ever got the chance to experience anything like this in their lives.

As if things weren't crazy enough I took on other side projects, often just for the fun of it. On top of the nGOmobile competition, we launched an SMS-based data collection tool called FrontlineForms in March 2009. This allowed very simple, low-cost phones to carry out structured data collection in places where there was no Internet, with the data neatly packaged up into multiple text messages, ready for sending to FrontlineSMS. We managed to squeeze in a little mobile gaming, too. Following the senseless murder of four rare mountain gorillas in the Democratic Republic of Congo in July 2007, we worked with Masabi and FFI to relaunch the Silverback game, hoping to raise renewed awareness of the plight of the species. News of that relaunch ended up being featured on the BBC World Service, Public Radio International, and on National Geographic.

All the media attention my work was getting was one thing, but the awards and Fellowships were something else altogether. When Stanford decided, albeit the second time around, to offer me my first Fellowship in 2006, it felt like something of a one-off. I didn't for one moment expect to receive anything like that ever again, but how wrong I'd turn out to be. Two years later Pop!Tech had come as a pleasant surprise, but then the floodgates really opened, with further awards coming in with surprising and unexpected regularity.

In 2009 I became a Tech Awards Laureate, a Silicon Valley badge of honour for innovators focussing on tech-for-good. That same year I was stunned to receive an Emerging Explorer award from National Geographic (more on that later). In 2011 I became an Ashoka Fellow, and FrontlineSMS won the Curry Stone Design Prize. As if that wasn't enough for one year, I won the Pizzigati Prize for Software in the Public Interest, and became a World Technology Network Fellow. More would come later, but only after my Imperial Phase had effectively come to an end following my departure from FrontlineSMS in the spring of 2012.

I was grateful back then to have a fiercely independent wife, no children and no hefty mortgage, and to not be rooted to any one spot. This freedom allowed me to take full advantage of all the travel and work opportunities that came my way, of which there were many. I can't remember turning down any interview, or speaking or writing opportunity during my five year Imperial Phase, none at all. I was enjoying absolutely everything I was doing, getting up in the morning with a spring in my step. I took nothing for granted and was mindful that it was only a few short years earlier that I had been lonely and frustrated, struggling to find purpose and meaning in my life. Now that I'd found it, and was living it, I squeezed it for every last drop and pursued everything with relentless enthusiasm and energy.

It was something of a miracle that I'd managed to do so much on my own, but I knew it couldn't continue forever. Thanks to the generosity of my donors, after four years I was finally able to bring in a little help, and the process of creating some order and structure around FrontlineSMS could begin. I hired people entirely on impulse, a risky strategy but one that paid off time and time again. Alex had been my first hire. After the relaunch I felt vulnerable promoting a platform I had little technical knowledge of, so when Alex reached out and asked if he could work full-time on FrontlineSMS, I jumped at the chance. After leading development of the platform at Masabi for two years, he finally joined us in November 2009 and, as you'd expect, he hit the ground running. Hiring Alex brought a critical function – platform development – in-house, which allowed us to build new FrontlineSMS releases, and fix bugs, without having to go through a process of negotiation with a third party each time. We hired another developer soon after, Morgan joining to provide backup and support. W+K, who were still deeply involved in our work, gave us free use of their office in the trendy Brick Lane area of London, and Alex and Morgan – and additional UK-based staff as our workforce grew – worked from there. I joined them once or twice a week, spending the rest of the time working from home in Cambridge.

As things continued to pick up it became obvious that I needed support myself. Laura came on board as project manager a few months after Alex, and took over much of the day-to-day managerial responsibility. Two months later came Ryan, who joined us to help with fundraising and partner development. We'd first met Ryan after Ben had contracted him to carry out a similar role at FrontlineSMS:Credit. Ryan would work remotely for us in Chicago. Flo joined us in January 2011 to manage community development, and Amy started a month later, managing the FrontlineSMS:Radio project. Finally, Sean joined in March 2011 after an initial introduction from Josh, and a year exploring his own SMS-based project ideas for the legal community. Although I'd given up responsibility for hiring some time earlier, it was a source of pride that we ended up with a roughly 50:50 gender split within the organisation.

It would have been easy to fill an entire book with all the incredible things that happened between 2007 and 2012, a phase in my life which I will forever look back on with happiness, gratitude and pride. Few people ever get the kind of opportunity I did, a chance to be intimately involved not only in the birth of a sector, but to build a technology that started life with so little, yet ended up achieving so much.

Seven-and-a-half years after coming up with the idea for FrontlineSMS, everything was in great shape, and it felt like the software was everywhere. People were desperate to use it, to build their own solutions on top of it, to fund it, and to come and work for the organisation behind it. We'd assembled a brilliant and dedicated team, had won more awards than we thought possible, and had users in 190 countries, many of whom were proud and enthusiastic contributors to one of the most vibrant of online communities. What's more, we were supporting organisations that, in turn, were providing humanitarian assistance to upwards of 30 million people. To top it all, hundreds of thousands of organisations had downloaded FrontlineSMS, an incredible achievement. We were well funded with the Omidyar Network, the philanthropic organisation created by eBay founder, Pierre Omidyar, joining the Rockefeller Foundation to support our work, the former to the tune of $1 million. In case further evidence of our rise were needed, in January 2012 we were named one of the Top 100 Global NGOs by Global Journal, and I was interviewed for, and featured in, a book about *100 Great Disruptive Heroes.*

Despite everything, I was beginning to feel it might be time to move on. I was happiest in the field, and now I was hardly ever there. I'd become increasingly removed from the users, and increasingly removed from their realities. It was my passion for people, and understanding them and the problems they faced, that drove me to build FrontlineSMS in the first place, yet I was now spending most of my time far away, running an organisation. I never wanted that at the beginning, and I certainly didn't want it now. Besides, there were other ideas I wanted to pursue. Never one to avoid the difficult decisions, in May 2012 I announced that I'd be stepping back from the project, and the organisation, that had given me so much. It was an announcement that caused something of a stir. FrontlineSMS had a healthy bank balance, and it was still very much on the up. Back then, founders didn't just up and leave like that.

Looking back, given where I'd been a few years earlier, FrontlineSMS saved me. Quite literally. Like a good friend, we'd come together, grown together, learnt together, and suffered together. Without it, I'd be nothing today. Saying goodbye wasn't easy, but it was the right thing to do for both of us. After a short transition period, Sean and Laura took over the running of FrontlineSMS and The kiwanja Foundation, and I was gone.

The story of what happened next would be someone else's to tell.

Live Curious

I doubt many of the students realised how lucky they were, let alone what an incredible honour it was. Most were still in nappies as South Africa's apartheid regime finally, thankfully, began to unravel in the early 1990s. The elderly gentleman now shuffling his way towards their table, breakfast tray held unsteadily in both hands as the ship gently rocked him from side to side, was one of the greats. You wouldn't have believed it now, watching this kind, smiling, peaceful, gentle character in a soft, flat, black cap and glasses, but for many years he was one of the few things that stood between hundreds of black protestors and a hail of police bullets.

I never did quite figure out how Archbishop Desmond Tutu ended up on that ship. I still have to pinch myself when I think about how I ended up on it, too. Events leading up to my arrival still feel random, crazy and a little surreal – and that was before I discovered there would be a Nobel Peace Prize winner on board. Daniel Epstein, the founder of the Unreasonable Institute, had tried many times over the years to fly me out to his base in Boulder, Colorado, to spend part of my summer mentoring the latest batch of Unreasonable Fellows. We never quite managed to make anything work, but he never gave up. A few months after I'd stepped back from FrontlineSMS he tried again, this time attaching a few photos of a glamorous-looking cruise ship, complete with beautiful cabins and even more beautiful sea views. My interest was piqued. 'I might just have got you this time, my friend', he wrote. And he had.

The chance to spend a few weeks effectively squatting on a cruise ship packed with Semester at Sea students, sailing around Asia, mentoring technology startups as we went, was just the kind of crazy idea that only Daniel could come up with, let alone pull off. 'Unreasonable at Sea is a radical experiment in global entrepreneurship to combat the greatest challenges of our time. 20 mentors. 100 days. 1 ship. 13 countries. 11 ventures. 1 belief that entrepreneurship will change the world.' The idea was so out there that I simply couldn't refuse. I wouldn't be able to do the whole three months, but I agreed to fly out and jump aboard in Hong Kong, stay with them for four weeks while we docked and held events in Vietnam, Singapore and Myanmar, and then fly home after we'd wrapped up our work in India.

I knew very little about the region, so getting the chance to see five countries in one go was something of a draw for me. Most of my work had, after all, been in and around Africa and, if I'm honest, Asia had never piqued my interest in quite the same way. Of all our stops, Myanmar fascinated me the most, though. It was a country that had been largely sealed off from the outside world thanks to years of strict military rule, so arriving there would feel like a step back in time. Just three years earlier, in 2010, a landmark election had signalled the end of almost half a century of army control. The election was considered problematic by many observers, and Myanmar was still finding its feet as a fledgling democracy by the time we arrived. It didn't last long, and a decade later the country would once again be in the news as the military had a change of heart and began wrestling power back.

Our ship, the MV Explorer, was owned (or more likely leased) by Semester at Sea, a US-based organisation known for its three-month, ship-based education voyages. A little over 500 paying students would spend one semester (or term) aboard, usually in the spring or autumn, circumnavigating the globe, learning and exploring as they went. The spring 2013 voyage would be a little different thanks to the addition of around a dozen tech startups and just shy of two dozen high-profile mentors. Fair play to them, Semester at Sea fully recognised the value of having such an elite group aboard, and the contribution a programme like Unreasonable could make to the students' overall educational experience. Daniel's pitch to them was probably a no-brainer, even if it was a little opportunistic and more than a little cheeky.

So, there we have it. One cold, dark February morning I left my wife and young son, Henry, with family in Finland and flew out from Helsinki to join Daniel and his team in Hong Kong. He'd assembled an incredible group of tech entrepreneurs, and an even more impressive group of mentors. Not for the first time I felt entirely out of place, and way out of my depth. We had Tom Chi, former Head of Product Experience at Google X, famous for his work on Google Glass and self-driving cars. And there was Meghan Smith, former Vice President at Google and former Chief Technology Officer for the

United States government, working under President Obama. As if that weren't enough, Daniel had convinced the founder of Wordpress, a Time Magazine Planetary Hero, the odd venture capitalist and a couple of Stanford professors to come along for the ride, too. There were no hangers-on here, just a beautifully assembled group of high-achieving C-suite stars, with centuries of experience at the highest level of business and government between them.

And, of course, me.

Thankfully, by now I was beginning to realise the value I brought to proceedings like this. My story was accessible, relatable, and made anything seem possible. It was the perfect pick-me-up for struggling startups, and the perfect message for students, or anyone, in fact, wanting to make a difference in the world but not knowing how, or knowing how but not having any resources to help get them started. Luckily for my fellow mentors, few had gone through the kind of protracted struggle and turmoil I had. Sure, they commanded a huge amount of respect, particularly from me, but they felt almost untouchable. I, on the other hand, was just an ordinary bloke who ended up doing something that happened to inspire and excite other ordinary people. This was a role I was becoming more and more happy to play.

I didn't have to wait long before I was given the chance to share my story with my fellow travellers. One night after dinner, shortly after lifting anchor in Hong Kong, Daniel and I sat down in the ship's main auditorium, an intimate space where all of the main events took place, and had a 'Fireside Chat'. During a well-attended evening, we spoke about our long-standing friendship, my early school years, my first time in Africa, FrontlineSMS, and my thoughts on the future of international development and social entrepreneurship. These were the kinds of intimate interaction I enjoyed the most and, going by the number of questions and the number of business cards I handed out at the end, many of the students enjoyed it, too.

Given how much I loved events like this, I was thrilled it wasn't the only one I'd be invited to take part in. And, it's fair to say, the biggest and the best would be kept until last.

Working with us on the voyage was Tori Hogan, a curious blend of author, filmmaker, travel photographer and global nomad. Tori had only just published her first book,

Beyond Good Intentions: A Journey Into the Realities of International Aid, which looked critically at the ineffectiveness of many of the refugee programmes she'd seen during her travels around East Africa. By now she was dedicating a significant amount of time and energy educating young people about aid, and how they might best focus their energies on making the world a better place.

During her daily exchanges with the students, Tori realised many were struggling with the poverty they were seeing in the countries we were visiting. While a few had managed to shrug it off, the vast majority were rightly beginning to question what they saw, and were wondering whether there was anything they could do. Figuring out how to help students grappling with these kinds of aid-related questions was right up Tori's street, and she checked in with Archbishop Tutu, or Arch as he was better known, and explained that she wanted his help. A few minutes later she'd pulled off something of a coup when he agreed to take part in a 'how to change the world' session for all the students on the ship. Given my own work, and my earlier 'Fireside Chat', Tori decided she wanted me to take part, too. As you can imagine, I jumped at the chance.

During the course of an hour, one midweek evening, Tori, Arch and I sat together in a packed, expectant auditorium, sharing ideas, thoughts and stories with a few hundred eager young students and faculty. 'How to change the world' was the theme, a phrase I wasn't particularly keen on, but probably the best, and easiest, way to describe what we set out to do. Tori directed the conversation brilliantly although, looking back, I wish she'd shared a little more of her own work that night.

Arch was everything I'd come to know him to be – thoughtful, hopeful, positive, humble, engaging and funny. At one point he tried (and failed abysmally) to pull off an impersonation of my English accent. I don't think anyone could quite figure out what he'd said, me included. After letting out the kind of full-bellied, infectious chuckle only he can, Arch rocked back in his chair, legs swinging high into the air, before reaching across and giving me a gentle fist bump. It was a simple, innocent yet wonderful gesture, and a highlight of all my interactions with him. And it was caught on video, too. Sometimes it was easy to forget who he was, and the incredible things he'd achieved in his life.

Luckily, it wasn't the only time we'd share a personal interaction like that on the trip. There was an ongoing joke aboard that, despite many of the passengers calling her a boat, the MV Explorer was, in fact, a ship. Rules dictated that anyone caught calling her a boat would have to do ten press-ups, on the spot, supervised, as punishment for their faux pas. One morning, during a brief conversation with Arch outside the dining room, it was my turn to slip up.

'Okay, Ken. Let's see you do it', said Archbishop Tutu playfully.

'Do what, exactly?'

He slapped me gently on the back. 'Ten press-ups, of course!'

I could hardly refuse, and the result was a photo of me, flat out on the floor, with the Archbishop looking over like a drill sergeant, counting and laughing, as I struggled through my punishment. He would later kindly record a short video for me, dedicated to our young son Henry, where he couldn't resist joking about what had happened. Despite his years, Arch never seemed to forget anything.

'Hello Henry Banks!' he bellowed. 'Welcome to this wonderful world. You have a great dad, you know. He doesn't know the difference between a ship and a boat! No, actually, I think I like him.' Henry and I often watch the video together. One day he'll realise how special that moment was, particularly now that Arch is no longer with us.

Arch and I didn't get many other chances to sit down and chat during the short time we had on the ship together – he was always in such demand, as you can imagine – but what we did get was enough to strike up a friendship, one which would inspire him to write the foreword to my first book, *The Rise of the Reluctant Innovator*, published 18 months later in November 2013. He also let me take a photo of him with his arms raised in the air, representing the W+K FrontlineSMS logo, which I shared with my former colleagues, much to their delight. We managed to keep in touch for the next few years, at least until his retirement (he had a habit of retiring and then un-retiring), by which time his family decided the best way to slow him down was to take over his email account, an action that did, unsurprisingly, have the desired effect. As a result it's been harder to keep in touch but, to this day, I remain grateful not just to have met, but also to have had the chance to work with, and get to know, such a wonderful human being. Arch was every bit as incredible as I expected him to be. We would meet one final time a couple of years later, in Oxford during Skoll World Forum, where he would sign a copy of the book we worked on together.

Unlike the chaos of my Imperial Phase, the years following FrontlineSMS took on a gentler pace. This 'Golden Age' would be dominated by far fewer but slightly longer, deeper, more meaningful engagements and experiences, of which Unreasonable at Sea was one. In a way, slowing down and picking my travels more wisely made sense after years of taking every single opportunity that came my way, regardless of how busy I was. Too often I'd take on trips not because I wanted to, but because it was in the best interests of the work, work that always seemed to come first. From now on, my wife and young children would come first, and I decided I'd only do things that were in either my or my family's best interests. After years of chaos the pace would drop noticeably, and the bar would be raised just that little bit higher.

It's not hard to imagine how spending time on a ship, sailing around Asia mentoring tech startups, with a Nobel Peace Prize winner on board, would meet that new standard, even if the opportunity wasn't in the slightest bit planned or expected. But by now Lady Luck seemed to be smiling on me, and she would play a big part in two other once-in-a-lifetime trips I would make during my Golden Age, one to Lake Como and the other, quite literally, around the world.

In 2012, at their annual conference, Pop!Tech and the Rockefeller Foundation announced an exciting new partnership, the Pop!Tech Bellagio Fellows Program, and I was in Camden Opera House to hear it. Bellagio, which I'd come across many times over the years, had always struck me as the most amazing place, professionally and spiritually. The thought of spending any amount of time in such incredibly beautiful surroundings, drowning in history, with some of the smartest people on the planet, discussing some of the world's biggest problems, caught my imagination. It felt like some sort of calling, but a calling very much out of my control. Bellagio wasn't the sort of place where you could just show up, after all.

Sitting pretty on the shores of Lake Como, Bellagio is a stunning village in the Lombardy region of northern Italy. Despite the number of tourists treading its quaint, cobbled streets or its calm, picturesque waterfront, it's managed to retain much of its early charm. It remains one of Italy's top tourist destinations, with a history of visitors going back centuries. Pliny the Younger was apparently one of the first, recording his visits in the first century AD. The weather is just about perfect, too, with year-round temperatures that rarely drop below 5 degrees Celsius, or rise above 28 degrees. The likes of Madonna, George Clooney and Richard Branson have all famously owned homes in the region.

None of this had anything to do with why I had my eyes on the place, though. Nestled halfway along the southern shore of the lake, on the tip of the peninsula, is the Bellagio Centre, an imposing complex of villas and historic buildings surrounded by 50 acres of beautifully tended gardens and sprawling parkland. Villa Serbelloni, as it was originally known, had been acquired by Ella Walker in 1930. Granddaughter of whisky distiller Hiram Walker, and widow of Prince Allessandro Della Torro e Tasso, Ella worried about what would happen to the villa as her health declined. With no heirs, she decided to look into donating the whole place, plus a two million dollar endowment to help with upkeep, to a US university or charitable foundation. After making enquiries with some help from the US Ambassador to Italy, the only person to show any real interest was the President of the Rockefeller Foundation, Dean Rusk. After much deliberation they were offered it, and took it. Ella died just one week after she signed over the property in 1959, taking comfort in the knowledge that the future of the villa had been secured – just.

What to do with the place was now the big question. While Rockefeller wrestled with a limited set of options, a number of distinguished guests decided to swing by, taking advantage of a rare opportunity to continue their work there. The Rockefeller Foundation decided to convene one or two gatherings of their own at the villa, too, and invited other non-profits to do the same. It turned out the answer was staring them right in the face – to use it to run residency programmes for distinguished scholars, policy makers, artists and musicians. So that's exactly what they did. Since then, around 5,000 individuals have at various times taken up residence at the Centre, including Pulitzer Prize winners and Nobel Laureates, and many more have attended conferences there.

The Pop!Tech Bellagio Fellows Program would bring together a small, interdisciplinary group of four to six hand-selected artists, scientists, designers, technologists and social innovators for a two-week immersion residency at the Centre. I was immediately sold on the idea, and desperate to be among them, so I quickly made enquiries, and even considered applying. 'I don't think you'll be a fit this year,' came the reply, 'but let's look at how things shape up next year'.

I thought my chance had gone but, as it turned out, 12 months later my luck was indeed in. The focus this time was on livelihoods, something I'd been taking a particular personal and professional interest in (I even gave a talk about it at Pop!Tech), and although I didn't make it as a Fellow, Leetha asked me to join as an advisor, or a 'catalyst' as they were known. I was ecstatic. Not surprisingly, the Fellows and other catalysts I'd be working with were the most incredibly smart, intelligent people. We'd gather in Bellagio for two weeks in August 2014, a location that turned out to be everything I ever imagined, and more.

Anyone taking a walk up the long, winding, at times steep cobbled main street in Bellagio town centre will eventually arrive at two large, imposing iron gates. That's as close as most people get to the villa, but it doesn't stop the odd would-be climber trying to get a glimpse of the famed building at the top. For those lucky enough to pass through the gates, a sweeping driveway takes you further uphill, leading on to the majestic villa itself. Small gravel footpaths branch off in every direction, an explorer's dream. Perfectly laid out flower beds border perfectly manicured lawns that sweep off into the distance, the view only broken by trees, stone statues and the shimmering blue of Lake Como itself. Two weeks would never be enough to discover every single pathway, every single route to every single tiny outpost, or outbuilding, that seemed to be scattered throughout all 50 acres of this majestic site.

I particularly enjoyed my night walks around the grounds, often alone but occasionally with the Fellows and other staff. Everything felt so different under cover of darkness. I trod the gravel paths regularly during my two-week stay, taking the time to think and reflect and be alone. It was dark enough to get lost, but most paths eventually led you back somewhere familiar. And it was on one of these evening walks that I first spotted what I assumed was a dimly lit chapel or church, a small patch of light on an otherwise vast, pitch-black hillside far across the lake. I was immediately captivated. What was it? How did it get there? How do *I* get there? I was still thinking about it the following morning when I sat down for breakfast with the Fellows.

'Have any of you been on the far side of the hill since we arrived?' I asked. One or two had but not at night, which was the only time that the small, glowing building would have been visible.

'It looks like there's some sort of church or chapel perched right in the middle of the hillside, across the lake. I was looking at it last night.' I didn't hold out a huge amount of hope, but decided to ask anyway. 'Does anyone fancy hiking up to it at the weekend, if I can find out how to get there? The views from there must be incredible.'

Nods of agreement all round. I was thrilled. We didn't know it yet, but that Saturday hike would be the highlight of our entire trip, not just for me but for all of us. It would turn out to be an adventure within an adventure, a sort of pilgrimage, a beam of light calling us over, and us being unable to resist. It reminded me of George Mallory who, when asked why he wanted to climb Mount Everest, replied 'Because it's there'. We were about to spend a day hiking to some mysterious building and back, basically because it was there.

It turned out that the light in the darkness was, indeed, a church. San Martino Church. Proudly perched about 500 metres above the village of Griante, the site began life as a small shrine before the church we see today was built in the 16th century, primarily to house a wooden statue of the Madonna and Child which had been found on the hillside. It's hard to explain – I've never been that religious – but it all felt so magical. The more I learnt about the place, the more I wanted to go there.

That night I plotted our route, which would begin with a pleasant boat trip across Lake Como followed by a three-hour hike through small villages, fields, patches of forest and open hillside. Dotted along the route were small shrines, and the odd chapel, each depicting Jesus at different stages of his life. I couldn't help but be in awe at the lengths to which people had gone, dragging materials hundreds of metres up a hill, to pay homage to their God. The trek was everything I hoped it would be, with beautiful sweeping views much of the way, and perfect, slightly overcast, breezy walking weather.

When we finally got to San Martino it was a struggle to take it all in. We ended up heading towards the gently sloping grass bank at the front, the majestic church to one side and spectacular panoramic views of Bellagio and Lake Como to the other. It was enough to take your breath away, and getting there turned out to be well worth the effort. I think we all got something out of that day, spiritually and physically, and it was the kind of opportunism, driven by sheer curiosity and a sense of adventure, that I loved.

So far, in a relatively short space of time, Lady Luck had spirited me onto a cruise ship with Archbishop Tutu, and given me a memorable two weeks in Bellagio. I'd have happily settled for that, but it turns out she'd been playing the long game, and she hadn't quite finished with me yet.

I didn't realise it at the time, but sitting in the audience during my very first appearance on the Pop!Tech stage in 2008 was a small group from National Geographic. As one of the sponsors of the event, they often used Camden as a way of keeping up with emerging ideas, and to help identify anyone doing what they considered to be interesting work in the humanitarian, social and environmental fields. The evening after my talk I had the chance to meet up with Alex, Cheryl, Anastasia and Sarah over drinks, an event we'd repeat with wonderful regularity over the next few years. They liked what I was doing with FrontlineSMS and, more importantly, we all got on wonderfully well. It was a huge honour to have anyone at National Geographic take any kind of interest in my work, but I had no idea where those early conversations would go. Always one to keep in touch with people whose company I liked, I made sure I popped into their offices to say hello whenever I found myself in Washington DC, which was often. The conversations were always fun, easy going and memorable.

Like many children growing up, I'd always been captivated by the mix of stunning imagery and exotic tales of faraway lands that lay behind the unmistakable, thick yellow border of *National Geographic* magazine. We could never afford to buy them ourselves back then, but we'd get bundles for next to nothing from jumble sales, and I'd often sit at home with scissors and glue in hand, cutting out my favourite bits for my scrapbook or bedroom wall. If there were any magazines left intact, I'd lie awake late at night with a torch, usually under the covers, staring wide-eyed as I thumbed through each page. For someone who sought escape, to dream, to think of other worlds, it was perfect. Sadly, for many of us feelings of excitement and exploration, of wonder and curiosity, often leave as we get older. Fortunately, they never have for me.

I knew National Geographic were interested in the work I was doing with mobile technology in developing regions of the world, but I had no idea how much they liked it. Of course, I wasn't privy to any of the conversations taking place over and above the ones I had with them, usually over coffee in one of the Caribou Coffees dotted around the streets outside their office. Not until one cold October evening in 2009, that is.

I was in London that evening after a hectic day of meetings, and was staying at a Travelodge up the road from London Kings Cross railway station ahead of an early start the next day. I remember checking in, taking the stairs to my room, and quickly opening my laptop to check my emails before heading out to grab some late dinner.

It was then that Cheryl's email arrived.

'CONFIDENTIAL

October 14, 2009

Dear Ken

Congratulations on being selected as a National Geographic Emerging Explorer for 2010!'

I have to be honest, I didn't believe it. One bit. This had to be spam, or some kind of prank, I remember thinking to myself. A slight typo in one of the dates in the main body of the message, talking about the news appearing in a *National Geographic* magazine that would have been published three months earlier, aroused my suspicions yet further. If that wasn't enough, I couldn't quite figure out what my work had to do with exploring. I closed my laptop and headed out to get something to eat, bringing a mushroom pizza and a pint of milk back to my room half an hour later.

Slice of warm pizza in hand, the TV quietly on in the background, I sat on the bed and pulled open my laptop. 'Right, let's have another look at that email.' The Emerging Explorer Program was genuine enough, and I could find explorer announcements going back years. I was still puzzled by the connection between my own work and exploration, though. Explorers from earlier years had been climbing, diving, scaling, digging or building, the kind of thing you might expect an explorer to do. How on earth did mobile technology fit in with any of that? (It turned out later that it does. Figuring out how to apply mobile technology in socially and environmentally meaningful ways *is*, they believed, a new kind of modern, 21st century exploration. It was a particular honour to have been the first mobile technology innovator to be picked as an Emerging Explorer. Other technologists, some of them friends, would come in the years that followed.)

To add to my overall state of confusion, there was no application process. Instead, nominations were made confidentially and anonymously. Who on earth would have nominated me? Other details in the email seemed legitimate enough, and I obviously knew Cheryl from previous conversations and meetings. As the lid gradually began to lift on the mystery, my mood changed from one of out-and-out disbelief to a more general kind of disbelief. This really was happening, wasn't it? An organisation I'd admired pretty much forever, whose programmes I'd watched and magazines I'd read in awe for years, and a place I'd dreamed of working since I was a child, was now recognising my work and offering to feature it in *National Geographic* magazine. Not only that, I was also being invited to their headquarters to rub shoulders with all sorts of amazing people to celebrate and share it. Thanks to FrontlineSMS I'd won a few things over the years, but nothing came close to this.

I did what I always did when any kind of good news came through. Leaving the remains of my pizza to go cold in its box, I phoned my mum. By this time she was quite ill, and I wanted her to continue to feel part of everything while she still could.

Despite my initial suspicion, Cheryl's email signaled the beginning of a long and productive relationship with National Geographic, one that would last for many years. My work was briefly featured in the June 2010 magazine, and a framed copy of that page hangs proudly on my wall at home. Not surprisingly, I sent a copy of the whole magazine to my mum. I had an incredible time talking about mobile technology and FrontlineSMS in their imposing auditorium, which on the day was packed with staff, accomplished explorers and Fellows attending that summer's annual symposium. I would randomly bump into people I'd only previously seen on television, or read about in *National Geographic* magazine, people like Bob Ballard, who discovered the wreck of the Titanic. (I had the honour of sitting next to him at dinner one night, and I sneaked his name tag home with me as a memento. I still have it safely tucked away in a box.) I also met Meave Leakey, world-famous paleontologist, and Michael Faye, who undertook that gruelling 15-month, 2,000-mile trek through the Congo Basin. I met new heroes, too, including John Francis, the 'Planetwalker', who spent over 20 years working his way across much of the Americas on foot, largely in silence. I have a signed copy of one of his books on my shelf, and we're still in touch today.

My talk at the Explorer's Symposium was followed, three months later, by another appearance, this time at Nat Geo Live!, a series of events open to the paying public that featured three or four explorers, usually held during the evening at their Washington

DC headquarters. Each time National Geographic invited me to do something my level of gratitude, and appreciation of them, rose several notches. Nat Geo Live! events were a big deal, a sign of confidence in me and my work, and in my ability to represent all that National Geographic stood for, and to communicate it in an engaging and exciting way to the general public. Posters were placed around Washington DC, and online, another surreal but proud experience for me. It really did feel like things couldn't get much better.

'Meet four gifted individuals recognised by National Geographic for making a difference early in their careers. This season's Emerging Explorers Salon, moderated by Benjamin Shaw, Executive Producer for the weekly radio talk show *National Geographic Weekend*, features transformative ideas that are influencing the world. Scientist Saleem H Ali promotes a pragmatic, inclusive form of environmentalism. Mobile technology innovator Ken Banks developed software that enhances the communications ability of people without access to the Internet. Agro-ecologist Jerry Glover uses biodiversity to improve food security. Activist Kakenya Ntaiya, in the face of daunting obstacles, founded the first primary school for girls in her region of Kenya.'

That night my nerves didn't get the better of me, and I had a brilliant time. Yet again I'd exceeded all my expectations, managing to bury all my worry, insecurity and self-doubt and deliver in front of several hundred people. A couple of days later I secured a coveted writing spot on the National Geographic website where, over the next five years, I'd feature over fifty projects and organisations from around the world using all manner of technologies to help make the world a better place. Opportunities to edit the series, called *Mobile Message*, became a priviledge I would share with others, too, inviting up-and-coming journalism or media students to guest edit articles, giving them something of a coup for their CVs in the process.

One year after my first appearance at the Explorers Symposium, I was invited back to help the 2011 Explorers settle in. My work was featured in *National Geographic Traveller*, the article accompanied by photos taken by a professional photographer a couple of months earlier in our lounge, and at the end of our garden, in Cambridge. I appeared on a National Geographic podcast, and my work was featured in a number of *Nat Geo Learning* biology textbooks to help children understand how FrontlineSMS was being used to track malaria outbreaks in South America. I was also asked to moderate the opening session of the Symposium the following year which, despite the pressure, was quite an honour and a tremendous amount of fun.

Sadly, I wasn't able to take advantage of every opportunity National Geographic put my way. Just a few months before our first son, Henry, was due, I was invited to take part in a once-in-a-lifetime expedition to Mount Everest. Alpine ecosystems the world over are under threat, victims of their own success, in a sense, as they attract unsustainable numbers of climbers and hikers. The Mountain Institute had been working over many years to protect what was left, and their latest project on Mount Everest was looking to deploy a communications network to educate climbers about the fragility of the environment around them, and share ways they could be more respectful as they travelled through it. Because FrontlineSMS had proven itself over and over again in these kinds of challenging, last mile environments, I was invited to join the team.

With a pregnant wife, and the obvious dangers of flying in and out of small, remote air strips on tiny planes, and potentially the odd spot of climbing, I reluctantly decided not to take up the offer. Dramatic as it may sound, the last thing I wanted was to have some kind of serious accident and leave my wife widowed, bringing up a child I'd never meet, all on her own. I'd always been up for adventure – a bit of danger, even – but it was now time to think of others, not just myself, and family would need to come first. Even if nothing did happen, me being stuck on a mountain tens of thousands of miles away from home was one stress my wife could do without. It would hardly have been easy to get home quickly if there was an emergency, either.

There were plenty of other things going on to be positive about, though. Over the next five years I continued to publish my *Mobile Message* series, and appeared twice more at the annual Explorers Symposium. I was featured as 'Explorer of the Week' where I shared, among other things, the last time I was lost, my favourite Explorer, and my favourite place on the planet (I forget my answers now). I was asked to sit as a judge on a flagship environmental innovation challenge, had an interview with Nat Geo Learning, and my work continued to be featured in revisions of their college textbooks. We have copies of those sitting proudly on a bookcase at home.

At the same time, I continued to meet up with Alex, Cheryl, Anastasia and Sarah at every opportunity, keen to learn more about what they were doing, keen to share what I was doing, keen to explore collaboration opportunities and, more importantly, keen for them to know how much I valued everything they had done, and continued to do, for me. None of the magic of those National Geographic years ever wore off, and none of it was ever, ever taken for granted.

Remarkably, despite everything, there was much more to come. With Unreasonable at Sea and Bellagio behind me, I thought I was done with my share of surprise adventures, but how wrong I was. It was now June 2016, and a typically hot, humid afternoon in Washington DC. I can't remember whether I was there on official National Geographic business, or just happened to be in town for something else, but I was asked to pop into their 17th Street headquarters to meet Ford Cochran, Director of Trip Talent for National Geographic Expeditions. Ford's office was a fairly small, low key, rather ordinary setting for a brief discussion that would have such magnitude, for me at least. I walked in and, after the usual greetings, pulled up a chair, still wondering what I'd been called in for. Ford kicked off by explaining that they'd been working with the *Wall Street Journal* on a new expedition that focused on innovation hotspots around the world.

As he spoke, I sat up a little. This sounded interesting.

'One of the destinations is Rwanda, and we know it's something of a digital hotspot, and we know you've done a fair bit of work across the continent.'

I shuffled forward a little more, edging closer towards the edge of my seat, careful to stop before slipping off. I knew what I *wanted* Ford to say next. The question was, would he? The silence between sentences seemed to drag on forever.

'So, we were wondering if you'd be interested in joining us for a couple of days in Rwanda, as one of our expert guides, to help our travellers understand what they're seeing, and why it matters?'

There. He said it. Wow! Really? An invite to Rwanda with National Geographic? I didn't need long to consider my answer, but before I could reply Ford jumped back

243

in. 'It goes without saying, we'll cover all your travel and accommodation, and pay you a daily rate to cover travel and work time.' Forget getting paid, I thought to myself. I'd do this for free. Of course, I didn't say that last part out loud.

I was over the moon. There's no other way to describe it. I left the meeting with a huge smile on my face and a draft itinerary of the whole trip, and the promise of a follow-up phone call in a few days. I was ecstatic to be involved, even if it was only for a short time. I sat down in a cafe near my hotel, ordered my regular latte and almond croissant, and meticulously worked my way through the paperwork. 'Encounter great feats of creativity and innovation on an expedition that spotlights human ingenuity – past, present, and future. From the startups of Silicon Valley and Barcelona to the pristine beaches of the Seychelles, join scientists, editors, and other thought leaders to discover how new technologies are changing the landscape in conservation, communication, transportation, and economics – and shaping the future of our planet.'

The expedition wasn't scheduled to depart California for over a year, in late October 2017, so there was plenty of time to prepare, or for things to change. And change they did. Not long after arriving back home I got that promised follow-up call from Ford. And it brought the most incredible news.

'Ken, we were wondering if you could join us for the whole three weeks, if you're free?' I didn't know what to say. Well, I did. I jumped at the chance.
'Yes, I'd love to!' came my reply.

The expedition was something of a first for National Geographic. They hadn't partnered with the *Wall Street Journal* on anything like this before, and I was honoured they were putting their faith in me. I'd never done anything like this before, either, and memories of all that earlier self-doubt came flooding back. Apart from Ford, I wouldn't know anyone on the expedition, and I didn't know a huge amount about the places we'd be visiting. Talk about leaving my comfort zone. Not only was this a leap into the unknown, but underperforming would *not* be an option.

One thing I did find slightly amusing, in the middle of all my worries, was the use of the word 'expedition' in the title. This wasn't going to be like any other expedition I'd been on, or any I'd seen on the National Geographic or Discovery channels – and it was certainly unlike anything any pioneering, early explorer would have experienced. This was 21st century 'expeditioning'. A converted Boeing 757 jet, its tightly-packed 233 economy seats ripped out to accommodate just 75 business class-style luxury leather seats, two across, would carry us and our paying guests, a dozen support staff, a chef, our very own doctor and a couple of expert guides. No slumming it here, either in the air, or on the ground if any of the hotels in the glittering brochure were anything to go by.

As one of the guides I was asked to give three on-board talks as we flew between locations, something I'd never done before, obviously. Delivering a presentation at 35,000 feet in a jet, with passengers spread down a long cabin, half of them out of view, isn't something I get asked to do often. Ever, even. As you might imagine, it was a bit of an ask but Rob and I (Rob was another, more experienced guide who also needed to present) made a pretty good go of it. It was the first chance many of my fellow travellers had to hear about my own work, and how technology was being used around the world to solve social and environmental problems. That was pretty much the focus of the expedition, after all.

The trip started well, with a wonderfully relaxing business class flight from London Heathrow to San Jose International, an airport conveniently located a short drive from Silicon Valley. I'd rendezvous with Ford, the logistics team and then, later, the guests, at the Rosewood Sand Hill, a luxury hotel located only a mile or two up the road from the car park at Stanford where I'd lived in my van ten years earlier. How far my life had come since then.

Any nerves I had about meeting the guests quickly disappeared at dinner that first evening, where a few of us – me included – stood up to help with general introductions and welcomes. If I'm honest, I didn't really know what to expect that night as I closed my hotel room door behind me, strolling through the stunning hotel grounds towards the main building. Would these be intense, high-flying, barely-retired, information-hungry, fussy, demanding former executives that I'd be dealing with for the next three weeks? Or relaxed older travellers, up for a little adventure with the odd question here and there, but on the whole just looking to get away, travel in luxury, and have a nice, occasionally intellectually stimulating break? I'd never hung out with the sort of people willing to spend $85,000 on a holiday before, let alone travel with any of them.

If I was coughing up that sort of money, I thought to myself, then I'd expect nothing but the very best.

Not for the first time, I needn't have worried. Without exception, they turned out to be the most incredible, wonderful, easy-going, kind, considerate people I could have wished to spend my time with. Forget the cliches, we *did* become one big happy family, and anything that might have resembled work very quickly felt nothing at all like work. The icing on the cake was my National Geographic colleagues, who were a dream. Ford was an ever-calming presence, making sure everyone was happy, both paying guest and non-paying guest. Then there was Susan, editor of *National Geographic* magazine. Calm, fun, intelligent and knowledgeable, she was everything and more you'd expect for someone with a one-in-a-million job like hers. Bob was our official photographer, and he and I had a brilliant time together. I've always been in awe of professional photographers, and they don't come much better than those whose photos get to appear in *National Geographic*. Bob had a great sense of humour, and we'd quite often end up sitting alone together, chatting over a beer when we first arrived in a new country. The photos he took on that trip, and the 30-minute video he recorded, edited and shared with us at the end, captured the whole adventure in ways few others could.

Finally, we had Rob, a 30-year veteran of the Society, and former senior editor of the magazine. Rob had been on many, many trips before and his friendship, support and advice made all the difference as I undertook my first. He was kind, thoughtful, totally dedicated to conservation and exploration (he'd done plenty himself over the years) and by the time we said our goodbyes on the steps of our hotel in Barcelona at the very end, we'd become friends. To this day, I still miss our time together.

You might expect an 18-day, round-the-world trip to be exhausting, hectic and over in a flash, but the opposite was true. Sure, we only spent about two days in each country, but flying direct each time, pampered on a luxury jet as we went, with no stopovers, and only once having to go through any kind of airport building (in security-conscious Israel, in case you were wondering) made all the difference. And so did the incredible efforts of the logistics team, who did all the heavy lifting for passengers and crew. Everything ran with military precision, and timings were planned to the minute, quite literally. I was constantly in awe of their attention to detail, and the brilliant and smooth way in which everything always seemed to run. There was little room for mistakes, which was fortunate given they never seemed to make any.

As soon as we landed, for example, we'd jump on a coach that would be waiting for us at the bottom of the aircraft steps, and be driven straight to our hotel. Forget little inconveniences such as having to actually go into an airport building, through passport control or customs, or standing around waiting for our bags. All of that was magically and expertly dealt with on our behalf. On arrival at the hotel we'd have some kind of welcoming party, one which usually involved large quantities of petals, singers, dancers, even elephants, and we'd all grab coffee, or a beer, or explore the grounds, while our rooms were quickly prepared (checking in was also taken care of). Once the rooms were ready we'd be handed a key and, on entry, our luggage would be waiting for us on the bed. Our room keys also came with a small envelope containing about $20 in local currency, just enough to get ourselves a coffee if we were out and about, or an ice cream, or a little souvenir. None of that messing around changing money like ordinary people.

I could get used to this, I remember thinking to myself, forgetting for one moment that all of this incredible service came at a cost.

When it was time to leave, everything happened in reverse. Before we headed out for our last dinner, we'd pack our main bags and leave them on our beds. By the time we got back they'd be gone, already making their way to the plane ahead of our morning departure. Basically, all the frustrating, tedious, boring, time consuming jobs were taken care of, leaving us all with nothing but the best kind of memories, and 18 days of beautiful, relaxing, stress-free travel.

After sampling the delights of Silicon Valley on the first full day, we headed out across the Pacific the next morning to Kyoto, Japan, stopping off to refuel in Anchorage and crossing the International Date Line, which lost us a day. After Japan we hopped over to Singapore before continuing our journey to Jaipur. I'd been to India a few times, but nothing prepared me for the majesty of Rambagh Palace, where we stayed, and the stunning red sandstone and marble of Amber Fort. Next stop was the Seychelles. I couldn't quite figure out what a tropical paradise had to do with human ingenuity or innovation, but thought it best not to say anything in case they changed their minds.

The two days we spent there felt more like a regular kind of holiday, although a luxury one at that, with sandswept beaches, clear blue sky, endless sunshine and panoramic views of wide open sea. The chance to scuba dive with Enric Sala was a highlight, too. A National Geographic Explorer-in-Residence, he quit academia to focus his efforts on saving the world's oceans after finding himself 'writing the obituary of ocean life', as he put it. Spending time with the likes of Enric is a rare honour. He's the kind of person you could ask quite literally any question about the sea, or ocean conservation, and get an immediate, detailed response. Who needs Alexa, or even the Internet, when you're in the company of someone like that?

Rwanda was our next stop – where I was originally meant to join the group – with an unforgettable journey up the Rwandan side of the Virunga Mountains to see the mountain gorillas. The weather that day was grey, damp and cool, and a deep, persistent mist hung in the air. The hike itself, through vast swathes of green, lush vegetation, was relatively easy on the legs, although it was a little longer than expected as our guides struggled to find the gorilla family we were meant to be tracking. Some of us began to worry that we'd never find them at all. But we finally did, spotting them in the distance first of all and then, gradually, getting closer as we gingerly edged our way in their direction. It was everything and more you'd expect from an interaction with these incredible, yet sadly endangered, creatures. We got surprisingly close, within metres in fact, in large part because they didn't seem to be the slightest bit bothered by our presence. As long as we remained quiet and didn't make any sudden movements, we were relatively safe. And, if any of the gorillas did get aggressive, or too close for comfort, the rangers were on hand and seemed pretty clued up with what to do.

While we stuck to the rules, the silverback had ideas of his own, launching into the odd charge just to remind us who was boss. On one occasion Bruce, one of the guests, ended up spread-eagled deep in the vegetation as he attempted, and almost failed, to get out of the way. The power, strength, speed and grace of these animals is a sight to see. Interactions of any kind, let alone this close up, rarely leave you. Indeed, many of today's conservation efforts can be put down to David Attenborough's unforgettable moment with a playful baby gorilla way back in 1978, captured beautifully on camera for the BBC's Life on Earth series. Those interactions stayed with him, too. Concerned that these magnificent creatures might be gone within a generation, and determined to do something about it, he convened a meeting at Fauna & Flora International, where he was Vice President, on his return to the UK. The Mountain Gorilla Project was the result. In time this would become the International Gorilla Conservation Programme, the organisation still tasked with protecting them today.

From Rwanda we headed west to Israel, where we'd visit Jerusalem's Old City and some of the country's oldest, holiest sites. Given obvious political and religious sensitivity, I was grateful that National Geographic had the foresight to invite both Israeli and Palestinian guides to travel with us. Hearing both sides to every story was refreshing, and reminded me how challenging so many of the issues were.

We arrived in Barcelona, our final destination, on the afternoon of November 8th. I'd been to the city a few times before, but never stayed in the kind of plush hotel we'd be staying in, and never been out with any expert tour guides before. It was fascinating to see the place in a new light, and to experience the incredible Sagrada Familia cathedral, Antoni Gaudi's final masterpiece, so huge in scale and ambition it's still under construction today, a hundred years after his death.

My final flight would take me home, via British Airways business class. It was a lonely one. It had been hard saying goodbye to so many wonderful people, and so many new friends. We'd all met and bonded in the most unusual of circumstances, the vastness and brilliance and randomness of the expedition making it all the more special and unique. We'd never all be together like this again. For me, those 18 days couldn't have been, or gone, any better, personally and professionally. I harboured dreams of being invited back on another trip, but knew it wouldn't be the same. The bar was now ridiculously high, and sometimes in life it's best just to leave it there.

Despite the highs of adventures like Unreasonable at Sea, Bellagio and National Geographic, everyday life after FrontlineSMS was also pretty good. I continued to receive invites to conferences and, although I didn't accept them all, I did enough to keep my hand in. I published two edited books sharing the work of innovators I'd either met, worked with or admired throughout my career and travels. The first, *The Rise of the Reluctant Innovator*, was self-published and came with that foreword by Arch. It did well, at one point becoming an Amazon 'Development Studies' best seller, and that led to an invitation to produce an updated, expanded version, this time with a professional publisher, Kogan Page. That book came with two forewords, one by

Ashoka founder Bill Drayton, considered the father of the social entrepreneurship movement, and the other by musician Peter Gabriel. It's one of life's little ironies that Peter agreed to write the foreword to that book. Life sometimes goes full circle, and his involvement bridged my two lives, the one I had before I found purpose, and the one I've had since. In the mid-1980s I often played his music as I battled with frustration and self-doubt, and listening to some of it today takes me right back to those lonely, uncertain days at home in Jersey.

How Peter Gabriel and I originally met is a story in itself. A world-famous musician and activist, Peter co-founded an organisation called WITNESS in 1989. WITNESS focused mainly on the use of cameras and other new technologies to record and document human rights abuses, and when mobile phones took off so did their interest. It was through this work that he learnt about what I was doing. Our first meaningful interaction was a rather unfortunate one, with me having to turn down an invitation to have breakfast with him in Barcelona after a Mobile World Congress event. I had a flight booked, and couldn't change it.

We did briefly meet in person at a Founders Forum event a few years later when I spotted him about to walk out of the hotel front door. I sprinted over, shouting (no doubt rather rudely) across reception as I went.

'Peter! Peter! So sorry to interrupt you, but I didn't want to miss you.'
'Oh, hi. Don't worry', he replied, placing his bag over his shoulder. 'I was just about to leave. You just caught me.'
'I'm Ken. Ken Banks.'
'Oh, hey Ken. Yes, FrontlineSMS right?'
'That's right. We've been in email contact a few times, and spoke about meeting up, but never managed it. So I just wanted to say hello, really.'
'Well, it's nice to finally put a face to a name. I'm a big admirer of your work. Let's connect again over email and see if we can fix up a time to meet properly, yeah?'
'Sure, thanks Peter', I replied, handing over my business card for no obvious reason. We shook hands, said goodbye, and he was gone.

We had some success with email, but never did manage to meet again. A couple of years after ambushing him at Founders Forum I decided to reach out again, this time asking if he'd be interested in writing the foreword to my second book. To my amazement I got an immediate reply. 'I'd be delighted', was his response, and true to his word he sent a draft over a couple of weeks later. That was a little over five years ago, and we still haven't sat down for a chat, despite our efforts. He did randomly ring me up one day though, which came as a bit of a shock. 'Hi, Ken. It's Peter Gabriel' isn't the sort of thing you hear often when you answer a call. I still hope to meet up with him, but the guy is pretty busy, as you'd expect.

Recognition, nominations and awards for my work didn't dry up when I stepped back from FrontlineSMS. If anything, they came in thicker and faster than before. I became a Fellow of the Royal Society of Arts, picked up a Cambridge News Special Award for Social Entrepreneurship (after skipping the black tie dinner thinking I wasn't going to win anything) and was nominated for the prestigious TED Prize. Fulfilling a life-long dream to visit Australia, I became a Visiting Fellow at RMIT University in Melbourne, and returned to San Francisco to pick up the Eugene L Lawler Award for Humanitarian Contributions within Computer Science. A little closer to home, I became a Fellow of Judge Business School in my home city of Cambridge. To top it all, I was appointed a member of the UK government's new Digital Advisory Board, became Entrepreneur-in-Residence at CARE UK, and was asked by the University of Sussex to be their Ambassador for International Development.

On the surface everything looked great. I was busy, able to support my growing family, still travelling when I needed (and wanted) to, and continued to receive welcome recognition from my peers. But as time went by I found myself becoming increasingly frustrated at the state of the global development and the social innovation sector I found myself in. I felt that too much money was being misspent, too many projects were poorly thought through, and fewer and fewer people were bothering to spend time trying to understand the communities they set out to help. Life for many 'innovators' had become too comfortable, and *feeling* as though they were doing good seemed more important than actually *doing* any.

Crucially, I was also finding less and less work that excited me, or that I believed in. Too often I was offered a contract or consultancy that went against some or all of my values, particularly my belief in simple, locally appropriate and locally empowering solutions. I didn't want to be one of those people who sold their principles down the river for any amount of money, however tempting it may have been (in some cases the salary or fee was eye-watering). I was finding myself turning against the sector I'd battled so long and hard to be a part of, and often went to bed angry and frustrated. I didn't want to be going to bed angry and frustrated, and after much deliberation I decided there was only one thing for it.

I'd never shied away from making the big decisions when I had to, and the one I was about to make would be the most difficult, disruptive and permanent one of all.

Losing My Religion

'Nobody can go back and start a new beginning, but anyone can start today and make a new end.' – Maria Robinson

It was short and sweet. Two sentences on my blog was all it took for me to sign off on an adventure that had run for an unlikely 33 years. Ironically, I never did fulfil my early ambition of a full-blown career in the humanitarian or conservation sectors. I may have dipped in and out many times, more through luck than good planning, but in all that time I was never fully employed by any single charity or global non-profit organisation. My belief that the only way to have a career with purpose was to work for one of these organisations proved wildly misguided.

In the end I did it my way, on my own terms, independent of the shackles, egos and politics that dominate much of the sector. I was free to be honest, to question and to be challenging, things that many others weren't able to do despite being frustrated by the very same issues as me. For a while, particularly during my Fauna & Flora years and, later, throughout my Imperial Phase, I was shielded from any pressure to do things I didn't like or agree with. But once I'd left the safety of FrontlineSMS, the curtains were slowly drawn open, and it would be only a matter of time before I became uncomfortable with everything their opening revealed.

I always saw my entry into the humanitarian and conservation worlds as a calling, not a career choice. Nothing I ever did over those years felt like a job, for which I feel lucky and immensely grateful. I was always driven to do things because, as a caring human being, it just felt right to do them. If I'd wanted a regular career I had a pretty good one lined up in the finance industry back home in Jersey, thank you very much, or at Cable & Wireless. But I hated the emptiness of it all, and soon realised that any kind of regular job was never going to cut it for me. It became painfully obvious, from an early working age, that my ideal job simply didn't exist, and that if I wanted one I'd just have to go out and create it myself. So that's precisely what I did. Young and idealistic, purpose was way more important to me than money, something I'd never really had much of anyway.

As things slowly picked up, one of my guiding principles was to only take on work that I believed in, and where I could add real value. I wanted to contribute genuinely to a project, not just visit a new country, climb a career ladder, earn good money or bolster my CV. It was a principled approach that always felt ethically sound, but it did come at a cost. Yes, there were plenty of opportunities up for grabs when I decided it was finally time to walk away, but for the most part I couldn't see where I could add any value to the work, or I found myself disagreeing with its aims. There was always something that didn't smell quite right, that didn't square with what I felt was the right way of doing things.

At the same time I was discovering that more and more organisations were nervous about hiring someone with a history of being outspoken and disruptive, someone who dared to challenge the status quo. Instead, they seemed more drawn to people willing to toe the line, to keep things as they were, to take the money, to keep quiet and not ruffle any feathers. That was never going to be me. Honesty may be the best approach, but it doesn't always pay off and, in my case as I neared the end, it was actively working against me.

Crucially, I decided I couldn't – I wouldn't – do anything that went against the approaches and principles I'd championed over the years. I didn't want my words to become hollow, and I didn't want to become a hypocrite, to sell myself down the river, to be a sell-out whenever it suited me. In the end I decided I'd rather have no work than the wrong work. The trouble was that, with a young family, lean spells as a consultant just didn't cut it any more. Something would have to give. The private sector would end up coming to my rescue, giving me a desperately-needed way out.

To make things worse, for far too long I foolishly believed I had the ability to bend the will of the humanitarian and conservation sectors in my direction, to steer them away from the waste, inefficiency and misguided approaches I saw repeated over and over again. We all wanted the same thing, to help the same people, and I thought they might be interested in what I had to say. After all, FrontlineSMS was a rare success story, and they'd surely be interested in what made it work when the majority of other people spent a lot more money than I did, only to fail spectacularly.

Wrong again.

But I never stopped trying. I published two books, both focusing on people who did extraordinary things with few resources, often outside the system, dedicating their lives to solving a problem they encountered that angered or upset them. Their stories ran counter to the multi-million and multi-billion dollar projects that dominated the pages of most books and annual reports. I wanted people to know that there was another way, a more gentle, respectful, slow and personal way of facilitating change, that they didn't need a job in a big organisation or charity in order to leave a positive dent in the world. If those books end up helping just a few people help a few hundred more, I'll be happy.

I published a Donors' Charter, a set of simple questions I felt donors should be asking before funding projects. The Charter was created out of a genuine desire to help them make better funding decisions. I figured that if the flow of money for poorly-planned projects could be cut, the result might be fewer poorly-planned projects. While a launch article I wrote for the Stanford Social Innovation Review broke site records for the level of engagement, with well over 100 comments generating a sometimes heated debate, nothing changed where it mattered – among donors, or those asking for their money.

I also launched a new initiative, Hacking Development. Billed as 'a four-part manifesto for change', Hacking Development focused on four things most people in the sector knew weren't working, in the hope they might be nudged to try things a little differently. Again, the idea was featured in an article in the Stanford Social Innovation Review and, again, not a single thing changed.

Failures aside, I was proud of my record proposing alternative solutions to problems, however unpalatable they turned out to be. Standing on the sidelines shouting how rubbish everything was wasn't my style and, to be honest, plenty of other people were doing that already. Sometimes, though, I was seriously tempted to join them.

The more I saw opportunity after opportunity squandered, with no obvious way of doing anything about it, the angrier I got. And I didn't want to be angry. I'd seen the transformative power of the Internet and mobile phones in places previously starved of communication and opportunity. I'd seen up close what happens when tools are made available that can be taken, owned and built on by people in those places, but hardly anyone else seemed interested in doing any of that. Building tools that were genuinely empowering effectively handed control over to local communities, and that just didn't fit in with the model. Communities were rarely consulted, and were trusted even less. The 'experts' needed to manage everything, from the money all the way down. The West knew best.

Yet, if you read the Twitter streams coming out of the many global technology-for-development events over the past few years, you'd have thought the opposite was true. Speaker after speaker extolling the virtues of working with local communities, empowering local actors, avoiding shiny new stuff and only using appropriate technologies, spending time 'on the ground' understanding the problems, and so on. I hate to say it, but the double-standards made me sick. Straplines, punchlines and soundbites were the order of the day, loved by Twitter but hated by almost everyone else with the ability to see right through them.

In the end, there really wasn't much more I could do except call time. While most of my work over the years was arguably on the fringes, I still felt very much a part of the wider global effort to rid the world of human suffering and environmental destruction. And, as part of that effort, I felt the frustration of its failures and shortcomings, whether or not they had anything to do with me, or whether or not I could do anything about them. After giving it my best shot it was now time to focus on what I could change, not dwell on what I couldn't.

Focus becomes ever more important with age, and I'm not as young as I used to be. Oliver Burkeman reminds us that we only have about 4,000 weeks on planet Earth and, put that way, it feels like an awfully short window of opportunity. It doesn't leave much time for arguments, getting angry, or doing anything other than being helpful, kind and productive in whatever positive way we can. The passion, empathy, insecurity and motivation to do good that I had in abundance as a child never did leave me, even if it is now wrapped up in an adult body.

None of these feelings takes the shine off what, for me, has been an incredibly rewarding journey, though. I ended up achieving far more than ever felt possible at the beginning, and my sense of disappointment in no way dampens those experiences, or the advice I believe I'm in a good place to share with others seeking purpose or meaning in their lives (you'll find that advice in the final part of this book). I may have taken my time, but I did find my purpose in the end, and I squeezed it for every last drop when I did. And despite my move back to the private sector, I never have lost sight of what matters. I strive for the same things, and a better world for everyone, regardless. I'm just coming at it from a different place.

Looking back, it would have been easy to come up with some sort of beautiful, happy, fairytale ending with me sailing off into the sunset, but from the very start I wanted this book to be a true and honest account of my journey, one that included the good as much as the bad, the positive as much as the negative, and the happy as much as the sad. While it may not be an exhaustive account of all the things that happened to me in that time, despite everything I believe my journey has been an incredibly positive one. Life rarely goes in a straight line, and rarely turns out as you expect, and that's certainly been true of mine.

Above, Sir David Attenborough and I share a joke during the launch of wildlive! at the Natural History Museum, December 2003. Below, with Twitter co-founder Jack Dorsey at a mobile activism event in Spain, March 2009.

Above, a rare photo of me in the field, working on a mobile phone
project evaluation in Malawi in 2016 (photo by Hayley Capp/CARE UK).
Below, the kitchen table in Finland, with the view of the forest outside,
where I wrote the first version of FrontlineSMS in summer 2005.

Above, my VW camper van, home for close to two years. Here,
parked up off Stock Farm Road on Stanford campus, March 2008.
Below, the BBC website shares news on the use of FrontlineSMS in
the April 2007 Nigerian presidential elections.

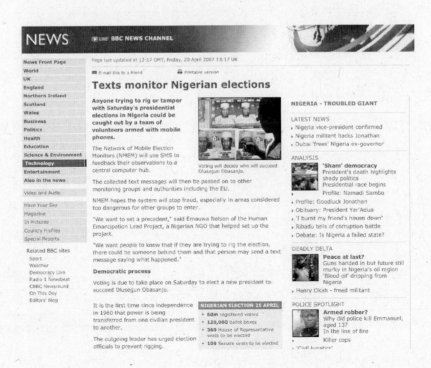

Above, relaxing on the banks of the Zambezi, with Zambia
off in the distance. Taken by Brenda during our visit to Mana
Pools, Zimbabwe, June 2006. Below, the SendConsole from the very
first version of FrontlineSMS written in Finland, August 2005.

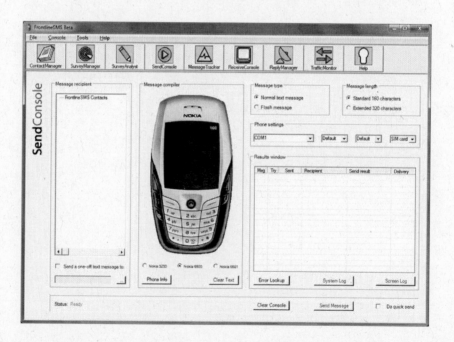

Above, speaking to students aboard the MV Explorer with Tori Hogan and Archbishop Desmond Tutu. (Photo by Evan Swinehart, Unreasonable at Sea, February 2013.) Below, preparing to give my big talk at National Geographic HQ at the 2010 Explorers Symposium, June 2010.

Joan Banks – nature's lady

Joan with a prize capture – a giant puffball Picture: NATIONAL TRUST

THE recent passing away of local botanist Joan Banks is as dreadfully sad as it is unbelievably unjust.

Sad that Jersey has lost a gentle lady with a huge and knowledgeable passion for all things wild and wonderful and unjust that her departure occurred during the season of spring that she enjoyed so very much.

Although Joan's reputation was largely based on her botanical interests, her knowledge of butterflies and fungi was equally recognised by scores of local enthusiasts.

As a much-respected member of the National Trust for Jersey's lands committee, her expertise was frequently imparted to the team of rangers who, within a comparatively short period of time, acquired a veritable cornucopia of tips and shortcuts in identifying all things bright and botanical.

Always quick to respond to any unnecessary manmade incursions into the natural environment, Joan was ever ready in giving strong and decisive views should certain land management undertakings not be quite to her liking.

Indeed, Joan Banks was one of those very special amateur naturalists who possessed the treasured skills of promoting much that we all see and enjoy in Jersey's glorious countryside – a lovely lady and a great loss to her family and everyone who knew and respected her.

The tribute to my mother, published in the *Jersey Evening Post* in the days following her death, April 2011 (reproduced with permission).

Above, a mountain gorilla strikes the perfect pose. Taken in the
Virunga Mountains during the Rwandan leg of my National Geographic round-
the-world trip, October 2017. Below, FrontlineSMS in use
at St Gabriel's Hospital in Malawi (photo by Josh Nesbit, 2008).

Above, Archbishop Tutu stands over, supervising my punishment for calling the MV Explorer a boat, February 2013. Below, San Martino Church in Bellagio, the 'light in the darkness', August 2014. Hiking here was a spiritual and physical highlight of our trip.

Part III

Purpose Explored

Culture Club

If you've made it this far and didn't skip Parts I and II, congratulations! It's an honour to have had the chance to share my journey with you. However you got here, I'm guessing you have a load of questions, some of which I may have answered already, and some of which I won't. It's the job of the final part of this book to help you make sense of what you've read, to challenge your thinking, and to help you plan a way forward in your own search for, or understanding of, purpose. You'll find plenty of advice which, I hope, will help you not only figure out whether it really is purpose you're after (it might be something else altogether), but also how you might go about finding it by yourself, with others in the workplace, or by developing and building ideas of your own.

But let's start with a few questions, and personal observations, to set the scene.

I've always been fascinated by people, although I'll never claim to really understand them. The more I've read, studied and written on the subject of purpose, the more confused I am about where the drive for meaning or purpose comes from, why some people seem to have it in spades while others don't have it at all, and how it might have evolved over time, in different places, among different people, and among different people in the same places. As for whether its origin is biological or cultural, that's an argument best left for the end of this book.

For now, let's go back in time and imagine your average prehistoric cave dweller, squatting in the darkness in loosely fitting animal skins beside a dimly-lit fire, belly half-full, eyes wide open, spear at the ready. With a life expectancy of little more than 30, most of their brief spell on earth would have been spent trying to avoid becoming a sabre-toothed cat's next dinner, or falling foul of some chronic disease. Dreams and ambitions beyond that, forget it. Their sole purpose would have been to stay alive and, if they were lucky, reproduce, just like most other living things on the planet at the time. The idea that life had any purpose other than survival and reproduction never really came into it. I suspect that, back then, cave kids never had the sort of conversations many children have with their parents these days about what they'd like to do when they're older. They'd just want to get older.

This line of thinking very much fits in with what Abraham Maslow describes as our 'Hierarchy of Needs'. According to Maslow, before humans can even consider things such as meaning, purpose, jobs, relationships or self-improvement, they need to have taken care of the essentials, things such as food, water, shelter, clothing and sleep. It's a well-cited theory, and one which makes a lot of sense. Sadly, it applies as much today as it did in prehistoric times, with billions of people alive today still struggling to meet even their most basic human needs.

Anthropology and curiosity are a potent mix, and I probably spend way too much time thinking about stuff like this. In my search for understanding I continually wonder where the concept of purpose came from, and under what conditions it first appeared. Purpose, at least in the context we're using here, hasn't always been a thing, after all, and if it likely didn't exist in prehistoric times and it does now, it must have appeared somewhere along the way. Why did it appear, and did it appear everywhere at once or as individual, random, isolated events? Does purpose mean different things in different cultural or geographical settings, assuming it's not the same for everyone everywhere? Is purpose to me the same as purpose to you? If you were born in another part of the world, would you think differently about it, or even care about it? And is 'purpose' even the best word to describe the thing so many of us seem to be looking for?

This book is, in part, an attempt to challenge you to try and answer some of these questions, or at least to encourage you to think more about them. I won't claim to have all the answers myself – believe me, I don't – but questioning things is an approach that's helped me a lot over the years, even if it has proved a little exhausting. Kids do it all the time. Adults should, too.

To start with, there's little doubt in my mind that where we grow up strongly influences our general sense of purpose. I often wonder whether my obsession with finding it is the result of my Western upbringing, for example. If I'd grown up on the other side of the world, would it have mattered as much? Or would it have mattered more? Looking at it another way, would I even have had the luxury to think about it if I'd been born to a poor family in a poor country? Purpose feels like a privilege when I think of all the places I've been, and seen, in my life. We all have dreams but, sadly, the energy we're able to put into pursuing them varies tremendously depending on the hand that life deals you. I guess I'm one of the lucky ones.

If I'm right and place is a factor, then what about religion? If I'd become a regular church-goer from a young age, would that have helped cement an early sense of meaning or purpose in my life just as it does, say, for monks or monastics? If it had, it would certainly have saved me an awful lot of pain and heartache trying to find my purpose years later. For some people, the answers they seek in life are very much rooted in their religion, and sometimes I feel like I've missed out. I admire people whose beliefs are so strong that they're willing to give up everything in pursuit of the purest of lives. Sometimes I wish Ray had been right and that God had found me.

When I started writing this book I figured that culture would have a key role to play in our awareness and perception of purpose. What I wasn't expecting was that religion and place would be just as important. I've always been open to the possibility that I sought purpose because everything in the West felt rather cold to me, rather clinical and mechanical, a life largely detached from the natural world and any sort of spiritual, greater meaning. In the context of wider Western life, while growing up I found that most people were more interested in climbing to the top and accumulating wealth and possessions as they went. That's what success supposedly looked like. Maybe you felt the same, depending on where you are today, where you grew up, or how you felt as you did. As a child I often felt in the minority when I wondered if life had real spiritual meaning or, in the words of Ralph Waldo Emerson, whether it would matter that I'd lived at all. I certainly hoped that it would. Not everyone on Five Oaks estate felt the same way though, probably because it was, well, Five Oaks estate.

Making sense of cultural interpretations of purpose (or meaning, or happiness, which are often bundled together with purpose) is a dream job for an anthropologist. If you're unfamiliar with anthropology, or find yourself struggling to get Indiana Jones out of your head, anthropology is 'the scientific study of the origin, the behaviour, and the physical, social, and cultural development of humans'. It's different from some of the other social sciences you may have heard of – such as sociology, perhaps – in three important ways. Firstly, it focuses on 'cultural relativity', meaning that an individuals' beliefs and activities should be interpreted in terms of his or her own culture, not anyone else's. Secondly, it calls for deep examination of context. In other words, the social and physical conditions under which people live. And thirdly, it focuses on cross-cultural comparison, in which one culture is compared to another. You may remember from earlier in the book that one of my first pieces of work at Sussex University looked at cultural interpretations of violence, and the dizzying array of definitions of violence – and justifications for it – among different groups and cultures around the world. Life is never boring, and never clear cut, particularly for an anthropologist.

All of this makes anthropology the perfect discipline for anyone wanting to study, understand, or make sense of the different views of purpose, happiness and life meaning around the world. Sadly, out of the thousands of books on purpose I've come across, very few seem to have been written by anthropologists. In fact, you can count the more general, non-academic ones on one hand. Why is that, I wonder?

One of the most studied, popularised and perhaps misunderstood cultural interpretations of life purpose has to be the Japanese concept of ikigai which, quite literally, means 'purpose of life' or 'reason for being'. You've probably heard of it. The Japanese believe that ikigai is something you possess within. The Western interpretation is that your ikigai sits at the intersection of what you're good at, what you love doing, and what's useful and needed in the world. Oh, and what you can get paid for. Mustn't forget that.

These are the main ingredients you'll read about in most ikigai books, and you'll find beautifully drawn, coloured Venn diagrams showing this convergence in most books on purpose, too, where they're often used as a roadmap to finding your purpose in life and work. Few of these books are written by Japanese authors, mind you, and remarkably there's very little English content written on the subject from those who understand and live the concept best.

Generally speaking, the Western interpretation goes something like this. If you're particularly good at something (say, writing) and you enjoy doing it, others get joy from reading it, and some are willing to pay for it, then you've got your ikigai. The key for many people here is probably going to be the 'getting paid' bit. I love writing, and it certainly sits in my sweet spot, but I'd be miles away from being able to make a living from it if I tried it full time. So writing is unlikely to be an option for me. Luckily, the Japanese don't agree with this interpretation, but more on that in a minute.

While it would be nice to find purpose in a full-time job, in reality few people do. In a 2010 study involving 2,000 Japanese men and women, only a third believed their work was their ikigai, or reason for being. What was most important was the feeling that their work, whatever it was, made a difference in other people's lives. Today, increasing numbers of young people are being drawn to roles within companies that communicate and act on a purpose that resonates with their own but, as Chuck Palahniuk, author of *Fight Club* reminds us, that doesn't make *you* your job. If you do find your true purpose at work, and you're getting paid for it, then you really are one of the lucky ones. The reality is, however, that most people are unhappy at work, and that's a real shame if you think about how much of your life you're going to be spending there.

If you take the original, non-Western interpretation of ikigai, you'll quickly realise that it's not some lofty, single objective at all, as some purpose books would have you believe. Instead, it's a way of being, often made up of lots of the small things that make life worth living. According to Ken Mogi, author of one of the few Japanese-authored books on ikigai, the Japanese 'do not need a grandiose motivational framework to keep going, but rely more on the little rituals in their daily routines'. Ikigai Coaches, an online community of practitioners, agrees. 'Ikigai is a multifaceted concept that Japanese come to understand as they live life and grow older. It is not something they learn about from a framework or from reading a motivational self-help book.' Ikigai turns out to be a state of being, not a destination or goal. When it comes to figuring out how to live your best life, the Japanese aren't as obsessed with goals as we are in the West.

The Ikigai Coaches also go on to say that ikigai doesn't have anything to do with making money, or doing something that the world needs. And if that weren't enough, you don't have to love what you do, either, let alone be good at it. In *The Little Book of Ikigai*, Ken Mogi argues that, instead, the five key pillars of Ikigai are: to remain curious and open, and to start small; to accept yourself for who you are and to drop any illusions; to live in harmony with the environment and people around you, and to live sustainably; to take joy in everything in life, particularly the small things; and to live in the here and now. Living in the moment is a theme we'll come across again and again.

There are plenty of ikigai resources out there if you're interested in digging a little deeper. Of the more prominent books by Japanese authors, check out Ken Mogi's *The Little Book of Ikigai* and *Ikigai – Giving Every Day Meaning and Joy* by Yukari Mitsuhashi.

Not surprisingly, the Japanese aren't alone in having a national concept of purpose, happiness or well-being. There are many other examples out there, and some have also broken into the mainstream over time as more and more people seek better, more sustainable, ethical, and spiritual ways of living. There's *hygge*, for example, the Danish concept of contentment. Hygge carries some similarities with one of the core principles of ikigai – to live in the moment. So, things as simple as lighting a fire or a candle, reading a book, taking a country walk or wrapping yourself in comfortable clothes would all count. The end result you're looking for is to feel calm, cozy, warm or content, or all four if you're really lucky.

There are plenty of others. The Germans have their own philosophy, preferring to combine different 'elements' of hygge into a single act. So, sitting in a cosy chair with a book and a hot drink, surrounded by loved ones and with music playing gently in the background is more their thing. They call this feeling of warmth and happiness *gemutlichkeit*. The Dutch have something that sits somewhere between the two, called *gezelligheid*. The Swedes have *mys*, the Norwegians have *fredagskos* and the Scots have *cosagach*, to name just a few. Despite originating in different countries, albeit ones rather close geographically, they all refer to similar concepts if you scratch just below the surface. It's not entirely clear how much any one may have influenced any other, but some of them clearly have given their considerable overlap.

The same can't be said for other life philosophies originating a little further afield. For example, *dharma* in India promotes living an honest, truthful, righteous life, and *guyub* in Indonesia encourages people to consider who they are within, as well as their place in, and connection to, wider humanity and society. Another popular philosophy, this time with roots in African culture, is *ubuntu*. One of Archbishop Tutu's favourites, it's a Bantu phrase that translates roughly as 'I am because we are', and it centres more on how your own sense of self is shaped by your relationships with other people.

Clearly, cultures the world over have gone out of their way to develop philosophies that help promote healthy living and a sense of being, both physically and spiritually. Learning to live and work well together is a vital ingredient for healthy societal growth and development, so it should come as no surprise that most communities develop and nurture philosophies based on contentment, respect and togetherness. And while some of these philosophies may have originated in a single, unique place without much or any external influence, others might have set root after arriving like seeds on the wind, carried from one place to another across land and sea as people travelled and the world opened up to trade and migration.

You could argue that finding contentment in life, or just in the moment, isn't the same as going out in search of its meaning or purpose. Many of today's lifestyle books have taken the easy route, choosing to focus on helping people make the best of what they have (or don't have), and supporting them in living their best life, whatever their circumstances. This is a pretty clever approach for authors wanting their books to appeal to the widest possible audience, encouraging them to believe that it's possible to make the change they seek. If the end result is happiness, contentment or a feeling of self-worth and well-being, then maybe it doesn't matter what you call it. I'd argue that setting out specifically to seek your purpose, or meaning in life, is far more challenging and potentially far more disruptive than seeking contentment in the life you already have. I could have saved myself a lot of trouble if I'd realised this before I started out.

So, perhaps now might be a good time to pause and reflect before throwing yourself head first into a search. Maybe you should ask yourself what you think it is that you're missing in your life, and why. I'll play Devil's Advocate here and ask whether it's really a lack of purpose driving you after all. Perhaps you already have everything you need to live a happy, fulfilling, purposeful life, and it just requires a little order or structure.

Simply having a set of rules to live by might be closer to what you're looking for, or need. Be open to all the possibilities because answers can often be found in the most unlikely of places.

It's clear that wherever and however society has evolved, some version of 'purpose' has evolved along with it, whether or not the people living there called it that. Most societies appear to have prioritised healthy living, and being a happy, collaborative, contented and valued member of society. This approach makes a lot of sense if we look back at our earlier cave dweller example. Rather than believing they were uniquely put on earth for some grand, spiritual, godly reason, these early people were more likely to adopt a worldview that was informed by how well they and their wider community coped from day to day. When life was brutal, getting by was all that really mattered. Things like having enough firewood, or food and water, or how safe the camp was, that's what would have been most important. It was more about the collective and less about the individual, an approach that may later have influenced communist thinking, perhaps.

Today, for some of us at least, life is less of a struggle. It feels like something of a luxury to have the time, freedom and resources to be able to choose to serve others, to choose how we spend our lives, or to think about seeking or achieving some higher meaning or lofty goal. Close to three billion people alive today have a pretty rubbish time of it, and many won't have the luxury or privilege to spend their time pondering life's bigger questions.

Don't lose sight of this as you embark on your own journey, and be grateful that you have at least had the opportunity to look, even if you don't end up finding anything.

$$X + Y = Z$$

So, I'm going to assume that by now, like me, you feel you've been put on this earth for a reason. And I'm going to assume that, again like me, you've been driven in some way to go out in search of it. Maybe you've found it – in which case, congratulations! Maybe you're just curious and, if that's the case, good on you. I'm a big fan of curiosity. And if you're still looking, don't worry. I hope to be able to help you get there, or get a little closer, at least.

If you got here via Parts I and II then you'll know that I went through an awful lot to find my purpose in life. Enough to turn into a book, in fact. It really ought not to take you that long, though, and I hope for your sake that it doesn't. That said, it's not a race, either, and for some of us the journey can turn out to be just as revealing and rewarding as the place we end up. The important thing is to do something, and to commit to taking a first step.

But before we begin, a quick reality check. Despite the many self-help books stacked high in airport bookshops telling you otherwise, there are no shortcuts and no magic formulas to finding your purpose. No book is going to do the hard work for you, and nor would you want it to. You likely already have the answers buried somewhere deep within. If I have just one single hope for this book, it's that it helps put you in the best possible position to have the awakening you're going to need in order to find it.

You'll remember that my awakening came in my late teens, with a global music event that opened my eyes to the vast inequalities in the world. Watching Live Aid unfold was only the start, though, and an uncomfortable one at that. What happened that warm July afternoon was a significant first step, but it was still just the beginning, and I struggled with what to do next. Donate a little money and move on? That's what a lot of people did, but that felt like a cop-out to me, and it hardly counted as making a difference or living out any kind of purpose. Images that genuinely haunt you don't disappear overnight, and the ones I saw that day certainly didn't leave me. Many of the people I've met and worked with over the years, at least those who found and lived their purpose, have images that never left them, either.

Although there were plenty of other people wanting to do good when I started out – clearly I wasn't the only one with a conscience – the route to helping was neither obvious nor easy. The quickest, most direct way I could come up with was to get a job in global development or, failing that, pretend to believe in God and join the church (the church happens to be very active in humanitarian relief). Given I had nothing concrete to offer the former, and I've never been drawn to the latter, I found myself somewhat adrift. Going it alone wasn't really an option back then, for all sorts of reasons. These were pre-World Wide Web and pre-mobile phone days, after all. Just imagine that for a minute.

Purpose, meaning and happiness weren't commercialised commodities when I was growing up, either, so there was no stack of self-help books to turn to when I felt I might be missing any of them. And because the Web was still a few years away, there was no Google to turn to. If that weren't bad enough I was also stuck on a small island, and there weren't that many like-minded people I could hang out with to share my frustrations. I was just left to stew. Eight years later I stumbled out of a pub and into a Jersey Overseas Aid project heading to Zambia, and things finally started to change.

If you're searching for some kind of purpose or meaning then things are very different for you today, but that doesn't necessarily make it any easier. It's still hard, but hard for different reasons. Helping people find their purpose has become big business and there are countless experts out there selling all manner of solutions left, right and centre. The problem has shifted from having *nowhere* to turn to not knowing *where* to turn. Too much choice can be worse than having no choice at all.

(Let me quickly state the obvious here. Yes, you're holding yet *another* book on the subject in your hands right now. What makes this one any different, or any better you may ask? Well, let me explain.)

In my first book, *The Rise of the Reluctant Innovator*, my aim was to challenge conventional wisdom in the world of social innovation. I found many of the academic, business and popular 'you can change the world' books at odds with my own experiences, and with those of my friends and colleagues who had successfully built solutions to complex, often nasty, problems. I struggled with the fact that many of the books available were written by people who had never built anything themselves, and I felt the personal insights gained by actually going through the process were missing. Similarly with this book, I hope that my own journey in seeking, finding and living purpose adds the kind of value that is missing elsewhere. I hope you find that to be the case, anyway.

Now might be a good time to have a quick look at some of those other books. If you've been brave enough to venture down to your local library or bookshop lately, you've probably noticed that much of today's self-help literature tends to gravitate around five key aspirations – purpose, happiness, meaning, success and getting rich – all things that most people will seek, ponder or wish for at some point in their lives. Life coaches and management consultants know it, too. Google has 275,000 search results if you're 'seeking happiness', 100,000 if you're 'seeking purpose', and 150,000 if you decide it's 'meaning' you're after. (I didn't bother searching for 'how to get rich'.) You could spend countless days, weeks and months watching any of the 30,000 videos made for your viewing pleasure, or enrol in online courses until they're coming out of your ears. Look yet further and you'll find hundreds of thousands of books with the single words happiness, purpose or meaning in the title. If you feel like you're missing *something* then how, exactly, are you meant to know which of these sites to visit, which of these books to buy, which of these courses to take, or which of these videos to watch? Good luck figuring that lot out.

Despite this mammoth library of resources, there's one thing almost all of these websites, videos, books and courses have in common. Look closely and you'll notice that most seem to use the words happiness, purpose and meaning interchangeably, as if they're loosely describing the same thing. Here's just one example. One of the few anthropology books I found on the subject had the title *Values of Happiness* but was subtitled *Toward an anthropology of purpose in life*. So, purpose or happiness? Maybe it's both. Perhaps if you find purpose in life, you also get happiness thrown in for good measure. Or, by extension, if you find happiness, you'll stumble across meaning. But then, you can be happy without purpose or meaning, right? No? I'm confused. You?

Many moons ago, Aristotle wrote that 'happiness is the purpose and meaning of life'. If he were to write a management book today he'd probably just bundle all of these concepts together, and not distinguish that much between the three of them. That's precisely what I was planning to do. For me, what matters most are the *emotions* we attach to the words, not the precise definitions of the words themselves. If you asked a few friends down your local pub, for example, most would tell you with some degree of confidence that there's some difference between the words purpose, happiness and meaning. But if you then asked them what that difference was, things would likely unravel quite quickly. Maybe I never did find my purpose. Perhaps my new-found work with mobile technology just made me happier, or gave my life new meaning. I couldn't honestly tell you anything other than that I feel I ended up in a good place. And maybe being in a good place is where most of us just want to be, however we get there, and whatever we call it. Maybe that's you, too.

Before we move on, let's quickly look up a couple of definitions, just in case your friends expect you to have an answer. According to dictionary.com, purpose is 'the reason for which something exists or is done, made, used' or 'an intended or desired result; end; aim; goal'. Wikipedia tells us that the meaning of life 'pertains to the significance of living or existence in general'. And the Oxford English Dictionary would lead you to believe that happiness is 'feeling or showing pleasure or contentment'. How these three things interplay in the real world, in our own minds and in the context of our own search, is what matters most. If you find it, you find it – whatever it's called – and you'll know you've found it. People who find their purpose rarely spend huge amounts of time analysing it, or wondering whether they really have found it or not. They just don't. Most people simply recognise that 'good place' when they find it. If *feels* right. It did for me, and it will for you, too.

Before we wrap things up here, let's forget purpose and meaning for a minute and just look at the concept of happiness. We're all familiar with happiness (or at least I hope we are) – that state of 'feeling pleasure or contentment' – but have you ever wondered if happiness is universal? Do you experience it, feel it or act it out the same way as everybody else? Does happiness to you feel the same as happiness to me? And does it matter, as long as it makes us feel good inside? A few years ago, Ana Swanson wrote an article for *The Washington Post* looking at meanings and expressions of happiness around the world. The star of the show was 86-year-old Mrs Xie, who found happiness in choosing the clothes she'd be buried in once she had died. 'Everybody does it!' she exclaimed. 'It's a happy thing', said another elderly lady. Just compare that to the sad and solemn feelings we have in many Western societies, where there is little happiness to be found in death.

You may by now be wondering whether all of this analysis is necessary. To be honest, I'm not usually one to dwell on semantics, but in the context of my journey, and in the context of where you might be in yours, I think it's worth the effort. It's worth it because more and more people are seeking purpose, meaning or happiness and, as we already know, there are thousands upon thousands of sources of advice on how you might find these things, all in the shape of videos, articles and books written by life coaches, academics, spiritualists and, increasingly, management consultants. Everyone seems to have an answer, an idea, an opinion, or a course of action promising to get you to where you think you need to be. Opening up your browser and searching the Web is not too dissimilar from walking into a shady, second-hand car showroom. Who, or what, do you believe when there's a potential sale on the line?

In the 20th Century Fox film, *Something About Mary*, one of the main characters, Ted, famously picks up a hitchhiker. During their conversation Ted's passenger shares his big idea, a '7-Minute Abs' video that's going to take the fitness world by storm.

'You heard of this thing, 8-Minute Abs?' asks the hitchhiker.

'Yeah, sure, 8-Minute Abs. The exercise video', replies Ted.

'Yeah, so this is going to blow that right out of the water. Listen to this! 7 ... Minute ... Abs!'

'Yes. Okay, all right. I see where you're going.'

'Think about it. You walk into a video store, you see 8-Minute Abs sittin' there, and there's 7-Minute Abs right beside it. Which one are you gonna pick, man?'

'Errr ... I guess I'd go for the 7', replies Ted.

'Bingo, man, bingo! 7-Minute Abs. And we guarantee just as good a workout as the 8-minute folk', emphasises the hitchhiker.

Ted gives an awkward glance. 'You guarantee it? That's ... how do you do that?' he asks.

'If you're not happy with the first 7 minutes, we're gonna send you the extra minute free. You see? That's it. That's our motto. That's where we're comin' from.'

'That's ... that's good', declares Ted. 'That's good. Unless, of course, somebody comes up with 6-Minute Abs. Then you're in trouble, huh?'

Things rapidly go downhill from there.

I share this because sometimes it feels as though many of the self-help books, life courses or videos are playing us with their own version of '7-Minute Abs', each offering a better, faster, easier, more comfortable or more direct way of finding purpose than their competitors. Anyone up for a copy of *Find Your Purpose in 24 Hours*? Or how about *Happiness on a Shoestring*? (If these books actually exist, no offence intended.) In truth, there's probably very little difference between many of them, content-wise, and who knows which one might be best for you. Many just package up and sell happiness, purpose or meaning as universal concepts, and they treat us – the customers – as though we're all the same. I think that's a mistake.

As I mentioned earlier, few of us really know if our own sense of purpose, meaning or happiness is the same as that of the person sitting or standing next to us. When I feel that my life has purpose, do I have the same sense of fulfilment that you do? When I'm feeling happy, do I have the same feeling of euphoria and positivity that you do? Or when I talk about meaning, am I talking about feeling valued in the same way that you do? Questions like these have troubled philosophers since the dawn of time, and I'm not sure we're any closer to finding the kind of universal answers that many people seek, self-help book authors included. Maybe it's because there simply are no universal answers, just a bunch of general rules and assumptions that can be applied in various ways. If that's the case, that makes finding the right advice even harder.

Humanists believe there's no specific, generic 'meaning of life'. Each of our lives has its own meaning. Interestingly, humanists also believe that meaning is not something you go out into the world looking for. Instead, they say, we should look at ourselves, and within ourselves, to discover what makes us happy and gives us meaning. Maybe purpose can be added to the list, too, although I admit I made the early mistake of thinking mine was somewhere 'out in the world'. In the end I didn't find it, despite years of searching, perhaps because it wasn't 'out there' waiting to be found in the first place. Instead, it ended up finding me.

So the very idea that a search for meaning, or a search for purpose, requires a search at all implies that the thing you're looking for might be out there somewhere when, most likely, it's within you and simply waiting to be unlocked, revealed, discovered, exposed or awoken. Sure, I jumped on my fair share of planes in pursuit of answers, and the events, people and places that enriched my life as a result certainly did me no harm. But everything I needed in order to turn that big reveal into reality lay within and certainly was not at university, or in the forests of Finland or Nigeria, even though I have no regrets about going to any of those places in search of it.

It's worth mentioning here that, while we might be driven to find answers, not everyone is, and it would be wrong to assume that everyone wants or needs a higher purpose. There's nothing wrong in feeling it's not for you, or in not feeling some need to dig deep, or travel the world in the hope of finding something. For many people it's enough to live life being kind, considerate and helpful. For others, survival or getting by might be all they wish for, or bringing up a healthy, happy family. Despite their often starkly conflicting origins, many religions share a common theme of kindness and compassion, believing them central to a good, fulfilling life. Kindness is a powerful and valuable trait, and if anyone feels that merely being a good person in life is a worthy purpose, then that's certainly good enough for me. But, just as some people seek higher levels of consciousness through religion, others may seek higher meaning through purpose. There's no right way or wrong way. You should simply follow whichever path feels best for you, or whichever one calls you – if there is a path at all – and not worry too much about what other people are saying or doing. We're all different, remember, so we should behave and make our own decisions as if we are. Only seek purpose, or meaning in your life, if you feel and hear it calling you, not because people say you should.

There's another popular school of thought worth mentioning here, one that claims the best way to find purpose is through your work. You just go out and find a job in a company with a mission or vision you believe in, and get paid for the pleasure of helping them deliver it. Bingo! Sounds easy, doesn't it? It's a popular line of thinking, but one that's being increasingly challenged, more recently in Charlotte Cramer's *The Purpose Myth*. I happen to agree that your purpose is highly unlikely to be your job (unless you create that job yourself, in which case it could well be). Sure, if you happen to work for a company that truly reflects your purpose, a company that not only lives it but lets *you* live it, then you're onto a winner. But few people do, unless you're self-employed and you've managed to build a business around something you love doing.

You can see why people try to seek purpose through work, though. Most of us will spend around half of our daily waking hours at work, and there's nothing wrong with wanting to make this time count. But piggybacking your life's purpose with your career is problematic on many levels, and can hand something you should be in control of over to someone, or something, else. What happens when you retire, or lose your job, or the company is taken over?

In a sense, thinking that your purpose is your job significantly increases your chances of believing you've somehow found it, and that could be one reason why so many books take that approach. It flips things around, focusing on the thing you spend most of your time doing, and *making* it your purpose rather than leading with the things you love and might *rather* be doing. Sure, this might sell books, but I'd argue it's a little misleading, and doesn't get to the heart of how we ought to be thinking about purpose in the first place.

One other problem with the 'your purpose is your job' mindset is that it rather cruelly discounts those who don't have jobs or, for whatever reason, can't work, or anyone who has retired, or anyone who volunteers their time (assuming that volunteering doesn't count as a job). Don't any of these people have purpose? For these reasons alone, I'm in the 'your purpose shouldn't be your job' camp. If you believe yours is – and you'll *feel* it, if that's the case – count yourself extremely lucky and make sure you never leave or get kicked out. If it isn't, as is the case for most people, then all is not lost. Instead, turn this to your advantage and consider leveraging any financial security it might give you to subsidise a side project that you dream of working on, one that feeds your purpose, or at least opens up opportunities that might help you find it. This is what I did for two years before FrontlineSMS took off in 2007, turning my part-time side project into a full-blown job that gave me more purpose than I ever thought possible. It's an approach that worked for me, and it can work for you, too.

If I could ask people just one question, it would be 'What could you be doing that would make you jump out of bed every morning for the rest of your life?' That's it. The question doesn't obsess on, or refer directly to, 'purpose' or 'meaning'. It's simple and direct, and anyone who leaps out of bed in the morning with a spring in his or her step is almost certainly ticking the 'happiness' box, so that's a bit of a bonus.

Agreed, I might be oversimplifying things a little. After all, some of those 'find your purpose' books run to hundreds of pages, so it must be more complicated than answering that one question, surely? Well, it doesn't have to be, and even the experts are able to narrow it down if they really want to. Here are three who did. To quote Christopher D Connors, Author & Executive Coach, 'If you have a natural talent for something, and you love doing it – you've hit gold'. Rebecca Sparks, who describes herself as a Confidence Coach, believes 'purpose aligns with your dreams, your values and your actions', while Jess Lively, a Life Coach, gives us a simple X+Y=Z: 'Your talent + Helping people = Purpose'. There are plenty more out there, but you get the idea.

Notice how each of these magic formulas list two main ingredients for purpose (I told you it was simple). They may not be the same two things, but two things nonetheless. Two is a pretty small number, but turning these handy little soundbites into something useful that works for you is the tricky part. So let me help you. To get the answer – *your* answer – experience has taught me that you need to do three things. For the sake of simplicity I'll call them the *finding phase*, the *matching phase* and the *actioning phase*.

Let's start with the **finding**. Grab yourself a pen and paper and think of all the things that excite you and interest you, things that you love, things that drive you mad, or anything that fascinates you. They need to be things that switch you on, make you sit up and pay attention, open your eyes, make the hair on the back of your neck stand up on end, make your heart race, make you angry, or make you cry. What you're looking to identify are things you can't get enough of, things you could bore your friends talking about (or complaining about) for hours on end. What do you read about more than anything else? What does your bookshelf say about you? What hobbies do you, or did you, have? What kind of TV programmes are you automatically drawn to? What would you do for a job if money were no object? These are all questions that only you can answer, and it's important that you're totally honest with yourself when you do. Start by making a list, maybe on a sheet of paper stuck on the wall, and add to it as you think of things that tick any of these boxes. They won't all come to you at once, and that's fine. Take your time, and see where you are after a few days. Go back and visit the list regularly, be brutally honest with yourself, and remember this exercise is all about quality, not quantity. A list with just a couple of solid items on it is probably better than one with dozens of average ones.

Next is the **matching** phase. Take a look at the things on your list and think about what skills, experience or talents you have that might be relevant or useful if you were to spend your time doing the things you are passionate about. It won't always be obvious, so you may need to think outside the box a little, or do some research to find a fit. And if you don't find anything, resign yourself to that fact, park it and move down to the next thing on your list. Something might click later when you least expect it.

I remember thinking to myself years ago, when I was keen to find a way into the world of conservation, that I had nothing to offer. I wasn't good at biology, and had no natural resource management, zoology, or wildlife skills to speak of. It turned out that none of that mattered. Most conservation organisations were crying out for people with technology and project management skills, things I did have. I ended up working on the frontlines of conservation in Nigeria for a year as a result, managing a primate sanctuary. So dig deep when you're thinking about matching your skills, experience and talents with the things you're passionate about. It will almost certainly require some research, but if you're passionate about it that part should be easy, fun and, hopefully, revealing. Take your time. You'll not get answers overnight.

Once you've matched one or more of your passions with one or more of your abilities, you're ready for the third and final step, the **actioning** phase. This is where you start reaching out into the world in search of tangible, practical opportunities to live out your purpose or passion. As before, opportunities or avenues might not always be obvious, so be open-minded, and make sure you read and research as much as possible. At the same time, keep circling back to the **finding** phase as you think of things you may have missed the first, second, third or even fourth time around. It's an exercise you'll hopefully only finish once you hit gold. Just remember, gold might not necessarily be a dream job, but perhaps a voluntary or part-time position somewhere, or a new hobby, or membership of an organisation or local group, or a place in an evening class or, as with me, a move to university. The **actioning** phase is not the end of the process, remember, but one which should put you on the right path to finding a purposeful or meaningful way of living your life. And as you go, focus on what *feels* right. Trust your instincts.

There's one final thing to remember that cuts across all these phases. Make sure you build up your networks as you go, particularly on platforms such as LinkedIn, Facebook and Twitter. Not only will it help you keep up with the latest news, but you'll be among the first to see new opportunities and, over time, these contacts can become the foundation of a very good support network. Get to know people, and let them get to know you. Share information about your search, and regularly. Social media may be a mess these days, but connecting you with other like-minded, passionate people is one of the few things it's still good for. It continues to work incredibly well for me, anyway.

Pay Attention

In the years following my departure from FrontlineSMS, I launched a number of initiatives as I grappled with how I might help other innovators and engaged citizens find an outlet for their talent, energy and passion for change. I figured the world needed as much help as it could get, and I didn't want to see any of their ambition go to waste. One of these new initiatives took the shape of a book – my first – called *The Rise of the Reluctant Innovator* which, as you know, sought to challenge many of the myths surrounding social innovation. That's the one that came with the foreword by Archbishop Desmond Tutu. Another was a website called 'Everyday Problems'. This idea took shape after I continually found myself in the company of young people desperate to find their purpose, or an outlet for their passion, and then use it to make a difference in the world. Many felt overwhelmed and had little idea where to start. Most didn't have any money, either, and many doubted they really had much to offer at all. In a sense, this is how most people, young and old, feel when they first set out on their search for a meaningful life.

'Everyday Problems' sought to lay bare many of the more widespread, universal problems in the world, such as child poverty, hunger and disease, rather than focus on obscure issues in remote places. Back then origin stories with a dash of romance or adventure were often the order of the day, and many achieved little more than to make social change feel unachievable, unrealistic or the exclusive domain of a lucky few. Purpose can sometimes feel like that, too. A strange tendency to hero worship the innovators themselves did little to level the playing field, either. I took it as my mission to democratise social innovation, to make it feel achievable to all, whoever they were, wherever they were, and however many (or few) resources they had. The same applies today with my efforts to help people with their search for purpose. Focusing on the more obvious, visible, everyday problems might not only increase opportunities for engagement and enlightenment, I thought, but might also provide an action plan that would be a simple and powerful way to help people get started and engaged. In a welcome twist of fate, the advice I used to give back then turns out to be pretty close to the advice I'd give anyone today.

The first action point is going to sound a little obvious, but it's one I've seen too many people skip over the years. It's to **understand**. Read widely, watch TV documentaries and other factual programming, and make an effort to make sense of the world and the problems and issues that concern you. Make sure you also take an interest in things that don't concern you, or concern you less. Inspiration can be found in the unlikeliest of places, and dots can join in the unlikeliest of ways. Trust me, without a decent level of understanding of how the world works, you'll not get far.

The second is a little less obvious than the first, but equally as important (and also often skipped in people's rush to go out and save the world). It's to **empathise**. Take time to understand what life is like for the people, or animals, suffering as a result of the cause or issue you're passionate about. If possible, spend time with them, ideally where they live, to best understand their reality. If you can't get there in person, see what you can do online, or connect directly with people who do have that lived experience. Second-hand knowledge beats having no knowledge, any day. Climb into the shoes or hooves of those you wish to help. Empathy is a superpower, and one in short supply. Make yours count.

My third recommendation is to **aim high**. By that I mean get behind a major global or national campaign that addresses one or more of the bigger issues you care about. Don't let the enormity of the task put you off, or the fact that you might never see the impact that you, individually, might have. Whenever I think about issues like climate change, I know there's no way I'll ever be able to point to any specific change that came about as a result of my own actions, despite doing everything I can. And that's fine, because I can take comfort in the knowledge that my contribution, added to those of millions of other like-minded people, will create the change we all want and need.

Look for volunteering opportunities with organisations tackling issues that bother you, either in-person or remotely. Become a member and recruit other members. Join demonstrations, sign petitions, stick up posters, give talks, start groups, get involved. This is one of the easiest first steps in seeking, or living out your passion or purpose, and it'll not only help broaden your understanding of the issues, but it will likely bring you into contact with people you'd otherwise never have met, or opportunities you'd never have come across. Think of it as a fantastic networking opportunity.

To achieve balance, and a sense of possibility, I suggest you also **aim low**. Get behind a local organisation or group addressing a local problem you're passionate about, or a smaller local 'version' of the wider global issue that concerns you. If that's climate change, for example, maybe help out with a sustainable transport scheme in your area. Get involved, dive in and personally experience the impact you're having. Resist the temptation to divert all your energies into problems far away. Draw comfort that you're making a difference to people and the planet where you live, and that you can see. The age-old mantra 'act local, think global' applies just as much today as it did when it first came into popular use decades ago.

Last but not least, **look to build**. In any spare time you have, use what you're seeing, hearing and learning to reimagine the world around you. What's not being done? What's being done, but not being done well? Be bold in your thinking, maintain a 'can do' attitude and don't let a lack of skills, resources or money get in your way. If you decide you need to build a website, for example, or learn how to use social media to share your ideas and thinking, then there are plenty of online courses out there, or books you can read, or YouTube videos you can watch. Learning to build something like an app might be a step too far, but even here there are plenty of sites where you can hire low-cost talent to build it for you. Some people will even work for free if they buy into what you're doing.

It might take a while to come up with a good project, and you may never do it, but be open to any opportunities that may arise. This is exactly what happened to me with FrontlineSMS, remember. I wasn't looking for anything at the time, but when the idea came to me I pooled my skills and resources to build and launch a platform which eventually took over my life, giving me the purpose I was desperate for. I didn't have heaps of money, or friends in high places. Trust me when I say this. If I can do it, anyone can. And starting with nothing makes success all the more rewarding.

Building and launching your own project is a big deal. It's hard work, and it will test your commitment. If you don't stick with it, it probably wasn't meant to be. I'll share a few more thoughts next on how you might best go about building something. It deserves a chapter all to itself.

When I look back and reflect, it seems my life was destined to be one defined by purpose and meaning. The earliest school report I have dates back to 1977. That would make me about 11 years old. At that age I was already being described as 'too emotional', a child who 'overthought things' and was 'a bit of a dreamer'. I remember being all of those things because they were all true. Most of them still are, over 40 years later. But, rather than seeing them as negative traits, as my teachers often did, I've learnt to see them as overwhelmingly positive. After all, these are the traits that make me, *me* – that make me care, make me cry, make me react, make me question, and make me act.

First and foremost, they're what make me pay attention. And only when you start to pay attention will your eyes begin to fully open to the endless possibilities around you.

Build It and They Will Come

So far in this book we've looked at how you might go about identifying purpose under your own steam, as I did, running off in whatever direction life takes you. In this chapter we'll look at two other options, starting with the idea of finding it at work. You'll remember that in the grand scheme of things this rarely happens, but it's certainly not impossible. If you decide to give it a go, I'll share a few thoughts on what you might want to look for in a potential employer. Then, we'll finish off with some tips on the final option, bringing your purpose to life by building your own organisation or solution – should you end up with an idea for one – that tackles a problem you're particularly passionate about. This was an approach, of course, that worked very well for me.

You'll remember how my 33-year adventure ended, but you'll probably not know what happened next. Well, I ended up back in the private sector, where it all began pre-Live Aid, working at a company called Yoti, a technology startup specialising in digital identity solutions (the kind that help you prove who you are). In 2018 I joined as their Head of Social Purpose, a fortunate title given the focus of this book.

It was quite a change for me, but an incredibly refreshing one. Most of my career has been in global conservation and development, a place where commitment to doing good is at the heart of everything. Even if the lack of progress is frustrating, the motivation to do good is unquestionable. The same can't be said for the corporate sector. Some companies do good as they go about their daily business, but many don't. Yoti, with its strong commitment to social purpose, turned out to be the perfect home for me.

By the time this book hits the shelves I'll not be far off four years into my new life back in the private sector. In that time I've learnt a lot about what purpose looks like, and doesn't look like, and what works and what doesn't. If you're interested in working for a company with a purpose other than just making money, or one which may help you live out yours, let me share a few things that I've learnt over the last couple of years.

Look for the badge of honour

We've all heard of 'greenwashing', efforts to cover up any environmental harm caused by a company's activities by producing a glossy brochure that pushes a few good things they might do. Anyone can do good PR or spin a good story, but not everyone can become what is known as a 'B Corp'. In the UK, B Corp certification is the badge of honour for companies looking to have a positive impact in the world. It indicates that a company balances purpose with profit. Stringently selected and heavily audited, these companies are committed to doing well by doing good. Yoti's B Corp status was a significant factor in my decision to jump aboard. Make sure you check out B Corps as you seek prospective employers, or any equivalent company structures in your country.

Find a company whose products you believe in

If you're looking to make a positive dent in the world, living out your purpose might be a little harder if you work for a company whose products don't improve the human condition in some way or another, or worse, have the potential to cause actual social or environmental harm. Look for a company with products that have the potential to do the kind of good you believe in. For me, this was a no-brainer at Yoti. Helping people to prove who they are, safely and securely, or keeping children safe online, are both big deals, and problems that aren't going to go away any time soon.

CSR doesn't equate to purpose

Companies can do great work through their Corporate Social Responsibility (CSR) programmes, but sectioning off the 'good things we do' into a separate CSR department tends to silo all the positives from the day-to-day drudge of the business. Most people want to work for companies that are good through and through, where everyone contributes, and not for companies that are about average with a few people running a cool little CSR programme in a corner somewhere. If you're thinking of joining a company, do some digging and see how embedded and inclusive their purpose really is, and whether the energy they put into it does actually result in some sort of positive impact.

Senior management buy-in is key

I've been fortunate at Yoti to have the full buy-in of the CEO, CFO and entire senior management team. Not only do social purpose programmes cost money but also, from time to time, they can divert resources away from other commercial activities that are crucial to the business. Building out a quality, meaningful social purpose programme without senior management buy-in is close to impossible, and gives a signal to staff further down that purpose isn't something the company genuinely believes in, or takes seriously. Purpose has to matter to everyone, from the top all the way down to the bottom. If the company you're looking to join has an impressive social purpose programme that runs throughout the business, and there's a specific role on the org chart for someone to lead and run it, that's a good indication of the buy-in you need.

Engage in a little silo busting

Yoti is a digital identity company, and our product offerings are somewhat niche (we're obviously working hard to change that). Rather than restricting outreach to those with a clear need for digital identity tools (something our sales team is obviously focused on), we've also been engaging the humanitarian sector (where our expertise has value) and various anthropology-focused networks (our social purpose work is very human-focused). We're also increasingly sharing our experiences with other people and organisations who work in social purpose, too, and in other industries. Think about the work you're either doing, or want to be doing, and imagine how it might cross over into other sectors and disciplines. I've often found this is where the magic (not to mention most of the out-of-the-box thinking) tends to happen.

Be evidence-based

When you're looking through a company's social purpose strategy, look to see if it's evidence-based. Doing things on a hunch sometimes works, and you can get away with anything if you put a little spin on it. At Yoti, a relatively small amount of money funded a piece of UK-based research during my first year which seriously challenged our social purpose ambitions. This led us to pivot to a more international strategy, one that you'll still see on the website today. Look for evidence that backs up, and justifies, the social purpose work any prospective employer might be engaged in. Doing it because it looks good is *not* good.

Embrace your 'known unknowns'

Reach out to your colleagues (or potential colleagues) and try to get a better sense of how they see social impact and social purpose. How do they define these concepts? What do their ideal social purpose activities look like? And how do they measure their impact? What do the media say about the company's activities? Honesty and transparency are key ingredients in any social purpose efforts, so struggles in this area might be an early warning of difficulties ahead.

Define the undefined

A couple of years ago we sent out a short two-minute survey to Yoti staff, and one of the questions challenged them to define our social purpose in one sentence. By way of an example, Disney defines its social purpose as to 'use our imaginations to bring happiness to millions', which I particularly like. The responses were fascinating and wide-ranging. Activities like this matter if you're looking for a single, coherent message to explain why your company exists, and to understand why other people feel it exists. Before you do much else, try to define what you mean by social purpose at work, and when you do, make sure that everyone supports it, or at least understands it.

Embrace the wider sector

At Yoti we don't just have eyes on our own products and services. We believe that a healthy digital identity sector is in the best interests of everyone. Because of this, a large part of our social purpose work is designed to support healthy debate around digital identity, and the democratisation of the technology behind it. We have invested time, money and resources into an exciting Fellowship Programme and the development of a toolkit, which benefits everybody, competitors included. Whatever the objectives of any social purpose efforts, don't forget to look beyond your own four walls. The problem is always bigger than any one company.

Build networks. Buy books

One of the most exciting things about starting something new is that it throws up the best possible opportunities to learn. Reach out to Heads of Social Purpose in other industries, and other countries. Ask them for advice. Ask them about their favourite social purpose books. Search for impact reports online to figure out how other companies define, measure and communicate why they exist and how the world is better because they do. These are still early days for corporate purpose, and you can help shape the future narrative as much as anyone.

Well, that's about it. Hopefully some of that helps if you're thinking about working for someone else as a way of living out your purpose. Let's look now at creating these sorts of opportunities for yourself.

If you're fortunate enough to have come up with an idea for a business, such as a tool, an app or a website to tackle a problem you care about, then you'll know how I felt that Saturday night way back in January 2005 when I had my big idea. Believe me, nothing beats building things yourself, and the learning you'll go through in the process. It's also the truest test of commitment, so before you dive in I suggest you take stock and think carefully about what you're doing, why and how you're doing it, and what the alternatives might be. Here are a few things to consider to help you get started.

Do you really understand the problem you're trying to solve?

You probably don't need reminding that I have a thing about people not taking the time to understand the problem they're hoping to solve, but I'll do it anyway. If you don't know enough about what's actually wrong, or why it's wrong, and why previous attempts to solve it have failed, then do yourself a favour. Pause, and do a bit more research, make a few calls, send a few emails. Believe me, trying to fix things that you don't understand rarely turns out well, and can often cause harm.

Are you the best person to solve the problem?

Put your own ambition and ego aside, be honest, and ask yourself why you think you're the best person to be solving this particular problem. If other people – perhaps those better qualified or with a better understanding or background – are already having a stab, consider bolstering their efforts instead. Don't reinvent the wheel. If it's impact you're after, sometimes you can have more by not building or starting something on your own.

Don't be competitive

It was a real surprise to find so many people with seriously competitive streaks during my years of humanitarian and conservation work. I somewhat naively believed that because we were all in the business of helping make the world a better place, we'd all play nicely together. Nope. Sure, some people do, but it's rarely a given. My advice is to try to be one of the good ones. Be respectful, be open, be responsive, and help people when and where you can. There's really no need to compete with anyone. As I used to remind people during my mentoring days, there's plenty of poverty to go around.

Don't be in a hurry

Overnight successes are rarely what they seem, and they rarely do happen overnight. Don't be fooled into rushing your idea to market. If it begins as a hobby, as mine did, take your time. Grow your idea or project on your own terms, and don't worry about being beaten by 'competitors'. Have faith that what you're building will stand on its own two feet, whatever else is out there. And, in case you're tempted, don't sell your car or home and throw yourself at it full-time until you're confident it has the potential to succeed. That's real, and often unnecessary, pressure to put yourself under.

Don't assume you need money to grow

I was always amazed by the number of people who, immediately after coming up with an idea, set off in search of funding before they did anything else. Sure, FrontlineSMS started with a modest amount of money, money that was needed to develop a working prototype of the solution itself. But after it was built and launched it was another two years before any significant funding came along. Don't assume you need money to grow or scale your idea. You can do an awful lot without any, as I did.

Do what you can before you reach out to funders

Funders like to support entrepreneurs, particularly entrepreneurs who show initiative. Those who reach out, having done as much as they can beforehand, have a better chance of success, in my experience, than those who show zero initiative and do nothing. Funders also want to know that their money is going to be spent wisely. Showing a little creativity before you get your hands on theirs gives them comfort that they're going to get more of a bang for their buck, and this is a plus in anyone's book or wallet. Run lean, even if you do end up with significant amounts of other people's money to spend. I always found it amusing that funders took comfort in my living in a van, but to them it showed commitment, and a willingness to run my life, and project, frugally.

Pursue and maximize every opportunity to promote your work

I remember how relentless I was when I first launched FrontlineSMS, using every chance I got to drop it into any guest post, presentation or interview I wrote or gave. See every conversation as a marketing opportunity, as a chance to get your idea onto as many screens and into as many minds as possible. However small they might be, they do add up, believe me.

Suppress your ego. Stay humble. Remain curious

You'll already know my thoughts on how destructive egos can be. Hard as it may sound, try not to make it all about you. Don't just share the voices of those your work empowers and supports, share your platforms with them, too. Help create spaces where they can share their stories, and connect and support each other, without you getting in the way. Remember, your success is intimately tied up with theirs, and that of any team you might build. Create a movement, not a cult of personality.

Get a good website

Remember that your website is, for most people, the primary window to you and your work. It's where many will get their first impression, and what they find there could be the difference between you getting that new user, interview, conference invite, investment or funding, and not getting it. You don't need to spend much these days. WordPress makes it relatively quick and easy to put together a nice, modern, clean-looking website. If you've never done it before, put a few days aside and learn. Being able to update and maintain your primary marketing channel quickly, yourself, will be important if your idea is to scale and grow.

Learn when to say 'no'

I've seen too many startups overstretch themselves early on, with some collapsing under the weight of mission creep. Make sure you manage expectations, and don't take on more than you can handle. Stick to your core mission or product, particularly when resources or funds are tight and, as difficult as it may be, learn when to say 'no'. Doing one thing really well is better than doing lots of things reasonably well, or poorly.

Avoid being dragged down by the politics of the industry you're in

Sadly in life, almost everywhere you turn you'll find politics. Over the years I've learnt to ignore most of it, choosing to focus my time and energy on the things that matter. Most of the critics, 'experts' and commentators I came across had never built anything themselves, or lacked any understanding of life for my users, so I wasn't worried too much about what many of them said. Whenever you read a crazy tweet, or Facebook post, or article about yours or someone else's work, that angers, frustrates or disappoints you, remind yourself that they likely don't have the experience or insights you do. If you feel an urge to reply or dive in, sleep on it first. More often than not you'll see things more clearly in the morning, and you'll end up avoiding the temptation to take the bait, or you'll respond more constructively. Try to save your energy for more important things, things like solving the problem that got you going in the first place. Your users, for one, will thank you for it.

Learn to do what you can't afford to pay other people to do

As you develop and grow your idea, you're going to end up having to do all sorts of things that you've never done before. Don't let that stop you. If you need a graphic or design, and don't have anyone else to do it, figure it out for yourself. The same goes for a website, or a marketing plan. Few things will be rocket science, and there are plenty of resources online to help you do almost anything these days. See any roadblocks as a chance for a little professional development.

Be open with the values that drive you

As FrontlineSMS began to take off, I remember sitting down one day to try and capture the values that drove and informed its development, and the values that drove me. As we brought more and more people on board, it felt important to have values to share with them, so they knew what was expected when they joined. The values I wrote down that day included things like putting the users first, promoting a can-do attitude, inspiring and respecting others, and remembering the bigger picture. When you feel ready, sit down and try to articulate yours. They're a great North Star, wherever you might be in your journey.

Collaborate if it's in the best interests of solving your problem, even if it's not in your own best interests

You'll remember the importance I placed on being problem-led, and you know that one of my values is always to remember the bigger picture. I can't overstate the importance of both of these approaches. Too often, opportunities to advance a solution (and alleviate suffering) aren't taken up because they're not in the best interests of one or both of the organisations involved. If you're genuinely focused on solving a problem, perhaps a problem that angers or frustrates you, then be prepared to do whatever it takes to help the world edge a little closer to eradicating it, even if it goes against your own best interests. Solving the problem should always come first.

Make full use of your networks – and remember that the benefits of being in them may not always be immediate
More likely than not, one day you'll step away from the thing you're building and focus on other things. Or you might need help with a big fundraising or investment round, or need to share news of a new release, or a job opening. Never underestimate the value of building a network as you grow. In the end, the people you meet, interact with and befriend will be the highlight of your career and often, at the end, they'll be all you have left. Use every opportunity to connect with like-minded people, even if the benefits might not be immediately obvious. One day they will be. And there are good friends to be made out there, too.

Don't lose sight of the bigger picture
We've mentioned this one before. Remember that whatever it is you're trying to solve, it's bigger than any one person or organisation. Another version of this might be 'don't sweat the small stuff'. Remember how you felt when you started out, and don't forget that you started out to solve a problem, not to have an argument. Don't give in to distractions, and don't forget all the suffering and injustice in the world. It doesn't go away just because you've taken your eyes off of it.

Finally, strive to be a good person and a role model to others
This one has always been incredibly important to me. I've tried to do everything in my life with integrity, to be a good, kind, honest person, to give others time and support where they've needed it, and to give back. More often than not this is what you'll be remembered for, not for anything you might have invented or built.

So, there you have it. A summary of what I'd consider to be the most important lessons I've learnt over the years. I hope some of it proves useful as you embark on your own journey. As for mine, we're not quite done yet. There's just enough time to take you on one last adventure, one with the unlikeliest of beginnings in a small, sleepy village on the south coast of England.

Legacy

I could hear what sounded like the distant chugging of a miniature steam engine as I made my final turn towards the house, a modest-looking residence with oddly-shaped tiles adorning the front which looked a little like inverted fish scales. Cattle grazed in the surrounding fields, and life seemed slow and easy compared to the bustling city centre I'd left behind in Brighton. This was Ringmer, a small village in East Sussex, and I was here to meet William Frederick Martin, my third cousin four times removed. This was a man I'd begun to admire immensely after finding him buried deep down in my family history.

I approached the front of the house and, drawn by the noise, took a turn down the side. I took a few steps and knocked firmly on the half-open workshop door. No answer. Maybe William couldn't hear me with all the machinery running inside, I thought, so I tried again a little more firmly this time. Second time lucky. He glanced over through clouds of rising steam, smiled and killed the engine, which came to a juddering halt. I could just about make him out as the air cleared. He was as tall as I had imagined him to be, with a fine, full beard typical of the times, and was smartly dressed as people always seemed to be in those days, at least in photos. As he approached I could tell that time had not been kind to him, and he looked older than his years.

He reached over the workbench and we shook hands.

'Hi, William. I'm so excited to get the chance to meet you. Thanks for taking time out and agreeing to meet up today.'

'No problem, Ken. It's really good to see you, too. And I needed a break. Tea?'

We headed off into the house, leaving the workshop in silence behind. The living room was dimly lit, but furnished as you'd expect for a house from the late 1800s. The fire was almost out, but let off enough of a gentle glow to give the room a warm, cosy feel. A large portrait of his father, William Brown Martin, hung proudly on the wall. We sat down at opposite sides of the fireplace, William taking his favourite armchair. Questions galore raced through my mind. There was so much I wanted to find out about him, and I knew this would be my one and only chance to get any answers.

'I love your workshop back there, William', I began. 'It's a real boy's dream to have a place to work and play like that. I must admit to feeling a little jealous.'

William carefully poured the tea. 'Well, I'm not as young as I used to be, Ken, but as a boy I certainly felt that way. My father and I would spend hours there', he said, glancing fondly at the portrait hanging above him. My eyes couldn't help but follow. 'I learnt a lot from my father and was encouraged to explore, build and tinker with anything that took my fancy. It was brilliant.'

I nodded in agreement, turning my attention back from the portrait to where William sat. I knew what he meant. I was fortunate that my mother had encouraged me in similar ways, although having anything close to a workshop would have been a distant dream.

'Your father must have been very proud of you, and you've done some pretty incredible things in your life, without a shadow of a doubt', I continued. 'You know, your curiosity and inventiveness really resonate with me, and you've become a bit of a hero of mine, if I'm honest. Have you got a favourite thing that you've built in your workshop back there? And what were you like as a child?'

William laughed. 'Oh, my. Where do I start?'

'Anywhere you like', I gestured, laughing along.

'Well, I didn't have much of an education, although the school I attended in nearby Lewes was one of the better ones in the area back then. Luckily for me it was pretty strong on mechanics and technology which, as you can imagine, are very important for any inventor.'

I nodded, leaning towards him as I put down my tea on the table between us. 'So you didn't really have any formal training, then? You just figured things out as you went?'

'Pretty much', he answered, rather proudly. 'I first started tinkering around back there when I was in my early teens. I'd spend as much time there as I could. I was very curious growing up and I taught myself to design and build all sorts of things, all through trial and error.'

I knew William had lived and worked through some pretty incredible times. Alexander Graham Bell had come up with the telephone in 1876. The light bulb was invented just three years later. The Second Industrial Revolution was well underway. I was as curious as ever to learn more about his own spirit of invention.

'What sort of things did you build then?'

'Well, there were lots of things, really. Over the years I built everything from a generator to working model windmills, which I managed to sell. I also made electric batteries, a telegraph device and – don't laugh – an automatic feeding device for pigeons. That was pretty good fun to test out in the garden. No birds got hurt, in case you were wondering!'

I laughed, despite him asking me not to. 'Oh, I'd loved to have seen that!'

Our free-flowing conversation continued, but one thing I was desperate to hear about was the bicycle. My patience eventually ran out.

'Sorry to interrupt, William, but you really need to tell me the story about the bike. I'm fascinated.'

'Ah, I wondered when you were going to ask about that', he grinned. 'It's quite a story. I built that in 1868 when I was eighteen. They tell me it was the first bicycle to be built in England. I didn't invent it, of course. I'd seen a sketch in a newspaper of one in France and decided to make one myself, mostly out of wood. I have a photo here. See? It wasn't the most comfortable ride, mind you, but at least it moved when I sat on it and peddled.'

William handed me an old photograph of his bike sitting proudly on its stand across the road from the house. It was exactly as I had imagined, a rough cut of the bikes that were to come, and one that resembled something close to the shape of a Penny Farthing, with one wheel larger than the other.

'That's fantastic, and such an amazing achievement. A wooden bike sounds pretty uncomfortable, though, not to mention heavy. Did you take it out on a test ride anywhere? What did people say when they saw it?'

'I did take it out, yes, and almost made it all the way to Brighton.'

I shuffled sideways in my chair, grateful that it was more comfortable than the seat on the bike. 'Wow, that's quite a way.'

William leaned towards me, arm outstretched as he took back his photo. 'Oh, it was about eleven miles, but it felt a lot further on that hard seat! I managed a pretty decent ten miles per hour, though, and people were constantly stopping me or shouting to me.' He put his hands around his mouth, making the shape of a megaphone. 'Hey, what's that you're riding?' they'd say. There was a huge amount of interest in it. That was a day I'll never forget.'

Building the first bicycle in England was an incredible achievement. Of course, I knew it wasn't his invention, but I did wonder if he'd done anything to protect his design, like filing a patent. That would have seemed like the obvious thing to do. William's head dropped slightly when I asked him this, his hand rising to rest on his chin. 'Sadly, no', he said, disappointingly. 'It would have cost us about £100 to file a patent, which was a lot of money. I spoke to my father about it, and we decided it wasn't worth it. Things may have turned out pretty different if we had.'

'I know. Such a shame really. Did you ever get any recognition for anything you built?'

'Well, I won first prize in a competition in Lewes for a working model steam engine I built. I won that the same year I built the bike, funnily enough.' Rising slightly out of his seat, he pointed towards a small, silver trophy sitting on a sideboard across the room.

I was pleased to see that he got rewarded for some of his efforts, but what struck me most was that he appeared to do most of his inventing while he was still quite young, in his teens, or while he was still at school. I wondered whether he went into engineering, or inventing, as a career. He could clearly have gone places if he had.

'Not really', he went on. 'I did what most people did and worked in the family business, becoming an apprentice carpenter for my father for a few years. I got itchy feet, though, and after my apprenticeship decided I needed to broaden my horizons, so I moved to London and worked as a joiner there for a while. That would have been around 1872. In those days, moving to London was a pretty big step for someone who came from such a rural part of England.'

I admired his bravery and ambition. It reminded me of the many times I'd sold up and moved on in search of greater opportunity. Maybe we weren't so different, after all.

'Did you stay in London for long?'

William shook his head. 'Just a few months', he sighed. 'Sadly my father passed away unexpectedly so I was called back to Ringmer to help run the family business. We started doing a lot more building work after I returned and we won all sorts of contracts over the next few years, building cottages and houses, roads, schools and even bridges. So things turned out quite well, really. The business became really successful and we ended up employing quite a few tradespeople.'

I'd read about some of the things William had built. I was curious, and asked him what his proudest piece of work was.

'I guess one of the things I'm best known for, and proudest of, is the tower of St Mary's church, up the road from here, which we finished in 1885. You'll see it today if you go and visit. In fact, there's a bit of a backstory to this. There had been a rumour in the village for years that a wooden tower used to stand on the site before a fire burnt it down. During our building we found evidence of a previous tower, with foundations, some charred remains and a few broken pieces of bell. It was a pretty exciting discovery, and confirmed what people had thought. I wrote an article about it for the Sussex Archaeological Society. It might still be available somewhere if you want to look for it.'

William's brush with archaeology was interesting given it's a subject that's always interested me, although I ended up studying anthropology not far down the road in Brighton, funnily enough. I was still curious about his inventing, though.

'With so much going on in the family business, did you find any time to continue working on other gadgets and inventions? You were obviously very good at it, and it was something you clearly enjoyed.'

He nodded enthusiastically. 'I did, in fact. Because I was now heavily involved in the business I thought it might be fun to build things to mechanise some of the manual tasks our tradespeople were having to do. So I designed and built things like steam-powered circular saw machines and moulding machines, which all came in real handy. I think this is one of the reasons we were able to get so much done over the years.' He reached behind him for a folder, taking out a small newspaper cutting. It was an advert for the family business. He read it aloud, proudly. 'W F Martin. Established 1750. Builder, Contractor and Surveyor. Steam wood works and saw mills, Ringmer, Sussex.'

The word 'prolific' kept springing to mind when I thought about all the things William had achieved in his life. I'd come to understand him as a man of many talents, and at various times he had been everything from an inventor and a builder to a poet and an antiquarian. If that wasn't enough, he also had musical talents, a subject to which we eventually turned.

'I've always felt very musical,' I said, 'but never did anything with it other than playing clarinet for a short while in the school band. Music played a role in your life too, didn't it?'

'Funnily enough, music is in our family's blood, so it's no surprise to hear you were in a band. I learnt piano from an early age. You might not believe this, but when I was about five or six I played piano to Finnish soldiers who had been captured fighting for the Russians during the Crimean War. They were being held in the prison down the road in Lewes. From about nine I also played the organ at St Mary's, and later became the church's organist and choirmaster, posts I held for thirty years. Oh, and I also composed a number of pieces of music. Some of it was performed publicly. I'm not sure if that's still available anywhere, but it may be worth a look.'

We finished our tea. The fire had by now been reduced to a few smouldering embers, a signal that our time together was coming to an end. It felt like the right time to ask how he felt about his life, and all the things he'd achieved, and perhaps about any regrets he may have had.

'So, what happened to you in later life, William, if you don't mind me asking?'

William paused for a moment, staring ahead in deep thought. The loud pop of a crackling ember quickly brought his attention back into the room. 'Of course I don't mind, Ken.' He leaned back, a look of resignation on his face. 'Sadly it didn't end too well for me, despite all my earlier successes. One of our bigger clients let us down in the late 1890s, and the family business went into steep decline. We were left with some pretty horrific debts. I tried everything I could, but I just couldn't rescue it. It was a hugely stressful time for me, and my health suffered terribly as a result.'

We both stood up and, almost instinctively, took a step forward, gently putting our arms around each other in a warm embrace.

'I'm so sorry to hear that, William', I said, reassuringly. 'You didn't deserve that to happen to you or the business. But the more I hear about what you achieved in your lifetime, the more I'm in awe of you. Understanding what inspired you and learning about how you lived your life has really helped me understand what drives me in mine. It's been so lovely getting to know you a little more. Thank you.'

We left the fading light of the living room and stepped into the bright autumn sunshine outside, saying our goodbyes as we slowly walked our way down the garden path. I shut the low, wooden gate behind me and, turning one last time as I crossed the road, caught one final glimpse of William as he stepped through his workshop door, back to the place where he was happiest.

You've probably guessed by now, but this meeting never did take place. But I so wish it had. William died on 25th April 1902, aged 52, a full 64 years before I was born. Learning a little more about him, and many other relatives, as I wrote this book has given me a sense of belonging, place and history that I can honestly say I've never really had before. It's also helped me to make sense of who I am and what drives me, and has brought answers to questions I've struggled with for most of my life. Marcus Garvey, a Jamaican political activist, was right when he said that 'people without the knowledge of their past history, origin and culture are like trees without roots'.

For most of my life I've been without roots. Growing up we rarely had any meaningful contact with my father's side of the family, and there seemed very little interest among the grown-ups in having any. Almost everything about him was a mystery. His premature death certainly didn't help, and I think there may have been skeletons in the cupboard, hidden for years, that some of them wanted kept there. As for my mother, we had regular summer visits from her parents to our home in Jersey, but beyond that everyone kept themselves to themselves, and there was little contact, something that really didn't matter that much to the younger me.

But I'm older now, the pandemic has been a time for deep self-reflection, and many of the questions I had over the years have bubbled their way back to the surface. I also have a young family of my own now, and I want them to know about the wonderfully rich history that might, some day, help them to figure out who they are, too. Learning more about family members who I never even knew existed just a couple of years ago has been such an enlightening experience and, without them, I can safely say this book would not be in your hands today. A story that felt incomplete for so long now feels, well, complete, and for the first time I can dream about meeting people like William because I know so much more about them and their lives.

In the final few years of her own life my mother had begun taking more of an interest in her family history, and she'd already been to Brighton and Ringmer where William and other relatives had lived. Perhaps, with time running out, she was looking for the same sense of place and belonging as I was. Whatever her motives, I've learnt so much more about who she was, and what drove her. In fact, I almost feel like I've got to know her all over again, and that makes me incredibly happy, grateful and proud.

It turns out that William's life, amazing as it was, was nothing out of the ordinary in the family. There were many more stories like his waiting to be uncovered, of ancestors who had taken all manner of risks and grabbed at every opportunity life gave them. I admired their sense of adventure, how they pushed to become their best selves. In a way, as I come to terms with my own journey and unlikely achievements, I take comfort in knowing that I'm not the only member of my family to do that, or to end up in a place I probably had no right to be.

Mark Miller Martin, William Frederick's uncle, was another. One of the few relatives my mother found in her search, Mark was born on 31st May 1831, in Ringmer. He was the seventh of an eye-watering 13 children. It's likely his middle name was given in recognition of his great grandfather, Henry 'Miller' Martin who, it turns out, was an actual miller. Growing up in a small village, Mark would have received little more than a basic education and, like most of the men in the family, he left school and took up carpentry. His father was a bit of a restless soul, trying his hand at many things over the years, everything from carpentry to farming to brickmaking.

Seeking a better life, his father made the bold move to emigrate to America in 1848 with his wife and six youngest children. They settled in Syracuse, New York State, and took up farming. They clearly sent back favourable reports because a couple of years later Mark and two of his other siblings decided to join them there. On the 26th of May 1851, just short of his 20th birthday, Mark arrived in New York on a ship called the *Falcon*, a journey that would have taken a not-insignificant six weeks or more. What felt like a life-changing move for me, from Jersey to Brighton in 1996, was clearly nothing in comparison.

With a background in carpentry, Mark went straight to work building railroad cars in Syracuse. He must have been unusually talented because, just two years later, he moved to Adrian in Michigan, where he was made foreman of the Michigan Southern Railroad car shop. In 1858 he moved to Litchfield, Illinois to supervise the construction of railway cars there (Litchfield was becoming something of a railway centre at the time). Later he was made Master Car Builder of the St Louis & Terre Haute Railroad, and he spent a few years living in Cochran, Indiana, as a Master Car Builder in charge of car shops for the Ohio & Mississippi Railroad Company. He was also briefly Superintendent of the Cincinnati Division of the Cincinnati, Hamilton & Dayton Railroad, and for a time lived in McComb City, Mississippi as Master Car Builder for another major railroad company.

Mark's life story is closely tied to the development of the American railroad system. From the 1830s to the 1860s there was a massive boom in railway building, and railroads replaced canals as the main method of transporting goods. They were cheaper, quicker and largely unaffected by winter weather. There were many different independent railroad companies serving mostly local areas and by 1860, in the industrial Northeast and agricultural Midwest, the rail network linked most major cities. Then came a westward expansion, particularly after the Civil War of 1861–1865 as the West was progressively populated. The first transcontinental railway was completed in 1869. Mark was working on the American railways at probably one of the most exciting times in its history.

Not surprisingly, he did very well financially. By 1895 he was able to buy both the Beach-Davis Bank, renamed as M M Martin & Co (later to become the Litchfield Bank and Trust Company), and the Litchfield Foundry. At the time of his death in 1901 he was believed to be the richest and most influential man in Litchfield, a man well respected by both his local community and the railroad world. Mark's story is very much one of a boy from a small Sussex village who made his mark as a successful car builder and businessman in the New World, where his descendants still live to this day. It's a story of travel, adventure and unlikely achievement that very much reflects my own.

It was an act of pure serendipity that I found any of my ancestors at all, yet more evidence, in my mind at least, that this book was meant to be. As the pandemic rapidly took hold, Mark and William's stories were still unknown to me but, driven by my ever-growing curiosity, I went online and found a genealogist in Ukraine, of all places, and paid him a little money to see what he could find. It was Bohdan Berezenko who first alerted me to a relative who'd worked at Bletchley Park with the codebreakers during the Second World War (more on him later) and, perhaps more crucially, to a living relative – a sixth cousin once removed, to be precise – with a vast amount of information on the family. It was this random connection with Roger Fenner that changed everything.

It turned out that Roger was a respected, award-winning educator and author in his own right, and a former Professor of Engineering Computation at Imperial College, London. His engineering background certainly made sense given everything I'd learnt about William and a number of other family members. Since retiring he'd been dedicating much of his spare time researching the Martin family, and he'd unearthed a huge amount of detailed material which, after just one video call in the summer of 2020, he kindly offered to share with me. Suddenly, out of nowhere, thanks to a random connection in Ukraine and the kindness of a relative I'd only just spoken to living in London, I had hundreds of pages of the most incredible material to hand. As I worked my way through it all, answers to some of my biggest questions started leaping off the pages, and I began to recognise more and more of myself. My passion and curiosity for how things worked, for writing, for travel and for the environment all felt less of a coincidence, and more a sign of something I was destined to do. More answers lay in the past than I had ever imagined possible, my purpose among them.

Writing is just one of the themes that goes back centuries. William Martin, another relative born in Brighton, in 1860, wrote one of the first books on the English patent system, and contributed a number of articles on the subject to Law Quarterly. He also wrote chapters for books on the history of Sussex and, inspired by his cousin's discovery of pottery kilns, wrote a paper titled 'A Forgotten Industry – Pottery of Ringmer' for the Sussex Archaeological Society. Another relative, Edward Alfred Martin, wrote many scientific books, papers and articles and edited a number of journals. The breadth and depth of his writing was breathtaking, covering everything from the formation of Brighton's cliffs to Sussex geology, coal mines under Surrey, sea erosion, Anglo-Saxon remains and the protection and preservation of plants. And let's not forget Henry Martin, the Mayor of Brighton who we met briefly earlier in the book. Henry wrote *The History of Brighton and Environs* in 1871, an original copy of which my mother managed to track down in a dusty old second-hand book shop. His book has stood the test of time, and is still available today. There's also evidence of another Henry Martin who became Editor-in-Chief of the Press Association (PA) in the 1950s, one more ancestor with a strong link to professional writing.

With all this new information to hand, the meticulously researched and typed project I wrote on the topic of oil when I was about 12 years old makes a lot more sense now. Flicking back through that crinkled, black, A5 folder today, I still find it rather strange that someone so young would attempt such a thing. It turns out I was simply doing what my ancestors had done before me. Later I would go on to run a successful blog on the kiwanja.net website, and contribute articles to the BBC, *The Guardian*, *Harvard Business Review* and many others before, of course, going on to publish a number of books of my own. And let's not forget all of my poetry, and the competitions I won as a child. All of this was clearly in my genes, even if I didn't realise it at the time.

Writing was just one of a number of threads I'm now grateful to be able to connect to my past. Another is a love for conservation and the environment. You'll probably remember that my mother was a prolific amateur naturalist, and she encouraged us all to take an interest in the natural world from a young age. Whenever her parents came over to visit we'd all spend days out bird watching, or studying plants and butterflies. There's little doubt that her passion came from her parents, and my own children are already showing a healthy interest in the natural world themselves. It's safe to say this is something which will live on long after my death, and that makes me very happy.

It's also inspiring to see how engaged many of my family members were in the early conservation movement. William, for example, was an active member of the Selborne Society, Britain's first national conservation body dedicated to 'the preservation of birds, plants and pleasant spaces'. Edward Alfred was President of the Croydon Natural History and Scientific Society and a Council Member of the Brighton and Hove Natural History Society. He also became a Fellow of the Geological Society in 1895. If that weren't enough, Edward and his wife were early supporters of an organisation that would later become the Royal Society for the Protection of Birds, better known as the RSPB. Other relatives also made their own contributions, many researching and publishing papers on the natural environment in and around the Brighton area. As with my writing, it's easy for me to see now how conservation pulled me like a magnet when I was unsure which direction to take my life.

Activism is another stand-out trait, with some of my relatives involved in the suffragette movement, some campaigning to protect habitats under threat, and others supporting the work of charities for the poor. Again, my sense of service and duty to others almost certainly comes from the work they started all those years ago, and it just seems to be too much of a coincidence to see it any other way. It's wonderful to think that FrontlineSMS, something invented in the true spirit of the Martin family, ended up supporting so many activists fighting for good in so many places. I always felt my story might be something of a one-off in the family. How wrong I was.

I can't help but think that moving to Brighton in 1996 to study at Sussex University was fate, given all the other places I could have ended up. The city, and surrounding area, turns out to be insanely rich in family history, most of which I was unaware of back then. But I had heard of Henry Martin, the former Mayor of Brighton, and had been told he was a great-great-grandfather of mine. Born in Steyning, Sussex in 1813, he led a life of incredible achievement. In his capacity as Mayor he officially opened the West Pier in 1865, and during his lifetime was a Chief Magistrate, Justice of the Peace, a published author, saddler and harness maker to the Queen and Royal Family, a Member of the Board of Guardians for the Poor, a town commissioner and Vice President of the Brighton Volunteer Fire Brigade. Fondly and widely known as 'The Father of Brighton', he died on 24th April 1885 after an eventful life filled with public service. Nearly a thousand London-Brighton and South Coast Railway staff lined his funeral route, and a number of Brighton shops shut their doors for the day out of respect.

I also can't help but wonder if it's fate that my accident in Nigeria in 2002 led me to Cambridge, where I still live today, another place with close family ties. William Martin, for example, attended Downing College in the 1880s, ending his studies with a Doctorate in Law. Another local relative came to light thanks to Bohdan, the Ukrainian researcher I'd hired early in my search. Dr Ronald Gray only died a few years ago, and I can't help but wonder how great it would have been to meet him.

Ron was a Life Fellow at Emmanuel College in Cambridge, spending over 30 years there as an author and lecturer on German literature, history and philosophy. He spent three months in Germany as a student in 1938 during the rise of the Third Reich, and saw and saluted Adolf Hitler, something he later regretted. Described by *The Times* as 'a brilliant Germanist' he was later called up to work alongside the codebreakers at Bletchley Park, where he was the first person to read, and then translate, the intercepted message from German High Command announcing 'Our shield and Fuhrer Adolf Hitler is dead'. Ron Gray's time at Bletchley Park is celebrated in their official Roll of Honour.

Despite struggling with history at school, I've become fascinated with World War II in recent years, and read as much about it as I can. It's amazing for me to think that an ancestor of mine played a key role in helping end the war, and that he played it at such a hallowed institution. Proud is a word that doesn't even come close to how it makes me feel.

Understanding how our ancestors influence us became something of an obsession the further I got into my reading and writing for this book. For me at least, it became impossible to ignore the dizzying array of traits and interests many of those I found seemed to share. But scientifically proving any of it? Well, you'd need to develop some sort of test for that, and the very idea of developing a genetic test to determine human potential is hugely problematic. The holocaust is just one example of the horrors that can unfold when ideas like these are taken too far. Tests aside, though, it is a topic of interest for many academics around the world.

The field of sociogenomics, as it's better known, is a relatively new discipline looking at how our ancestors might impact our intelligence, or chances of success in life, and some of their early research does suggest a link. One study of nearly 1,000 people in New Zealand concluded that success was, indeed, shaped by our genes and another, this time with a sample size of over 20,000 people from the UK, US and New Zealand, came to the same conclusion that 'those with certain genetic variations earned more money, had better careers and got further in education'. Research featured by the BBC also found approximately 'two-thirds of differences in school achievement can be explained by differences in children's genes'.

You may not discover the same connections in your past that were waiting there for me, but simply knowing about your ancestors can have a hugely positive effect on your life. Known by researchers as the 'ancestor effect', just thinking about them can apparently increase your self-esteem. Chakell Wardleigh, a magazine editor, also cites resilience, a sense of identity and increased happiness as positive effects. She also believes that 'your ancestors stories can shape you into a more grateful, happy, empathetic and compassionate version of yourself'.

Throughout history, my family have put considerable emphasis on learning, curiosity, music, writing, engineering and a passion for the environment. These are all traits that appear again and again, right up to the present day. My own children are already developing the same interests that my mother and I had because my mother brought me up surrounded and inspired by them. And it's likely my children will go on to do the same with their own children, too. And so the cycle repeats.

But success isn't guaranteed. Looking back at the considerable achievements of many of her own ancestors, I can't help but feel our mother never did meet her full potential. It's clear she gave up a huge amount to marry my father, and then to bring the four of us up alone after he died. She always told us we were the best thing to ever happen to her, but her life was much harder than it needed to be, something that always troubled and upset me. It was only when I was offered my first major piece of work with wildlive! in 2003 that I was finally able to start supporting her financially, buying her a new car and sending her money to spend on herself each month. This was something I continued to do right up until her death. For so many reasons it felt like the right thing for me to do, and she thrived in her later years until illness cruelly took her down.

There was a surprising amount we didn't know about our mother, but I did know she'd done very well at school, took a good job working in *The Times* newspaper's head office in London, and was a natural when it came to understanding nature and the environment. But a rebellious streak saw her leave home young to marry my father, who she'd met in the Don Inn, a town bar, during a family holiday to Jersey in the early 1960s. They say history often repeats, and not only had her own grandmother turned her back on convention and married 'outside of her social class', but many of the girls in our family since have shown similar determined, rebellious streaks. It's yet another family trait, I think. Maybe I'll need to keep an eye on Maddie, my own daughter, as she gets older.

From a personal perspective the evidence is overwhelming, and I can't see how my ancestors didn't shape much of who I am today. In some way, given everything I know about them now, I've lived my life as they, and destiny, intended, and my purpose and reason for being were determined long before I was born. Perhaps if all those years ago I'd sought answers within my own family and not in faraway lands, I'd have realised that innovating, writing, conservation and serving others were my purpose, and the things I was socially, and perhaps genetically, programmed to do. I ended up doing them anyway, and that's what I find most incredible about all of this – that despite doggedly following my own path for much of my life, I somehow ended up joining my ancestors on theirs.

A studio portrait of 18-year-old William Frederick Martin, taken in 1868, the year he built what is believed to be the first bicycle in England. (Photo courtesy of Roger Fenner.)

Above, a photo of William and his bicycle, taken outside the family home in Ringmer, Sussex, in 1897. Below, the first bike ride described in Peter Ambrose's 1974 book, *The Quiet Revolution*. (Photos courtesy of Roger Fenner.)

2e The first pedalled bicycle ride in Britain began at these gates and proceeded uncertainly away down the road. The date was 1868 and the machine was made by W. F. Martin, who lived here. Note the tile hung and weather-boarded walls – both local features.

A sketch of Henry Martin, Mayor of Brighton, published around 1884 in the
Brightonian newspaper, a short-lived publication that featured individuals
considered to be playing prominent roles in local life. Note that Henry is
holding a copy of his own book on the history of Brighton. (Photo from an
original copy of the print owned by the author.)

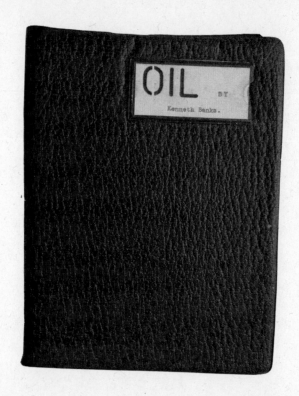

Above, the front cover of my oil project, written and typed on an old Olympus typewriter my mother bought me when I was around 12 years old. Sadly, my folder of poems from the same era has been lost to time. Below, the death of our father recorded in the parish register in Jersey (bottom entry).

3561	24ᵗʰ March 1972	29ᵗʰ March 1972	Annie Le Lievre	74	N.G. P. Rows N°6 St Peter's Ch. brand sto	St Peter	EVans Coleman Rector
3562 (see 2946)	2ⁿᵈ April 1972	4ᵗʰ April 1972	Irenee Madeleine Battam	58	N.G. SECTION III Row 15 · N°68 Ch. brand sto	St Helier	E.H.Dentith Vicar of St Luke
3563	3ʳᵈ April 1972	10ᵗʰ April 1972	Harry Banks	47	N.G. NEW GRADE SECTION XII Row 26 N°14 Ch. brand sto	St Saviour	EVans Coleman Rector

Above, Ron Gray in uniform during the war (image courtesy of his incomplete autobiography, available online). Below, how the world reacted to the death of Adolf Hitler. Ron famously translated the message from German High Command announcing his death. Image courtesy *The New York Times*.

Our mother's final resting place, a beautiful, sweeping, rocky bay
near the northwestern tip of Jersey, and one of her favourite places
when she was alive.